Pvt. Ltd.

Vegetable Varieties of India

H.P. Singh

Indian Council of Agricultural Research,
Krishi Anusandhan Bhawan II, Pusa,
New Delhi 110 012

Mathura Rai **Sudhakar Pandey** **Sanjeev Kumar**

Indian Institute of Vegetable Research,
P.O. Jakhini-Shahanshahpur,
Varanasi 221 305, U.P.

2009

Studium Press (India) Pvt. Ltd.

Vegetable Varieties of India

ISBN: 978-81-907577-6-8

Published by:

Studium Press (India) Pvt. Ltd.
4735/22, 2nd Floor, Prakash Deep Building
(Near Delhi Medical Association),
Ansari Road, Darya Ganj, New Delhi-110 002
Tel.: 23240257, 65150447; Fax: 91-11-23240273;
jngovil@studiumpress.com; shray@studiumpress.com

Printed at:

Salasar Imaging Systems
Delhi - 110035 (India)

Foreword

Vegetables, a rich source of minerals, vitamins, phytochemicals, antioxidants are integral part of human diet and there is an increasing demand with growth in income, and consciousness for health food. In the last two decade there has been an appreciable growth in production due to improved cultivars and technologies. In this context, Indian Council of Agricultural Research through the institutes and centres of AICRP on vegetables located at 57 centres has played a pivotal role. During the period, a total of 366 varieties/hybrids have been developed. This includes 174 varieties/hybrids in Solanaceous, 47 in Leguminous, 30 in Cole crops, 24 in bulb vegetables, 56 in Cucurbitaceous, 23 in okra and 4 in other vegetables. Information on these cultivars are scattered, resultantly desired information have not been available to the stakeholders. I am happy to note that the information on varieties, released for cultivation, has been compile together and is being brought out as book "Vegetable Varieties of India".

I compliment the authors for their efforts in bringing these useful information together, which will be of great value to all those who are interested in the improvement of vegetable production.

(MANGALA RAI)
Secretary,
Department of Agricultural Research & Education
and
Director General,
Indian Council of Agricultural Research,
Ministry of Agriculture, Krishi Bhavan,
New Delhi - 110 001

Dated the 5th February, 2009
New Delhi

Preface

Vegetables play a major role in Indian agriculture by providing food, nutritional and economic security and more importantly producing higher returns per unit area. In recent past, Indian agriculture has witnessed tremendous progress in vegetable production, due to increased production and productivity. The current area of 7.49 million hectares under cultivation and production of 116 million tons of vegetables amply justify the overall growth. The production of tomato and cabbage has increased manifolds due to extensive cultivation of improved varieties/hybrids. Now it is not only a question of providing enough vegetables for a balanced diet to more than one billion populations but also to provide quality vegetables acceptable in international market.

Since its inception in 1971, All Indian Coordinated Research Project on Vegetable Crops (AICRP-VC) has played proactive role in formulation and execution of various research and development programmes on vegetable crops. One of the major activities of AICRP-VC is to identify vegetable varieties/hybrids for its release and cultivation in various agroclimatic zones through multi-location testing. In recent past, hybrids are becoming more popular than open pollinated varieties. Nevertheless, in a number of vegetable crops open pollinated varieties continue their role to serve marginal and resource-poor farmers.

A large number of high yielding and disease resistant varieties have been identified and released. Despite the tremendous importance of vegetables, no literature is available on the vegetable varieties/ hybrids released through Central Varietal Release Committee or identified from AICRP-VC, which cover information and description of general interest of the varieties. Majority of vegetable growers are largely unaware of the improved varieties/hybrids. Thus it was felt a need of hour to compile the details of each vegetable variety/ hybrid released or identified for release. For a better understanding of each crop, in addition to the vernacular and botanical names, a general description has also been given, which will provide necessary background to the readers. In addition, a list of vegetable varieties either released through Central Varietal Release Committee, or identified through AICRP-VC has also been given.

We are very much thankful to researchers from different ICAR institutions, state agricultural universities, and private seed companies, for the excellent cooperation extended in supplying the desired information, and willingly sharing the relevant photographs.

We express our deep sense of gratitude to Dr. Mangala Rai (Director General, ICAR and Secretary DARE) for encouragement, motivation and guidance for preparation of this book for the benefit of vegetable breeders, farmers and students.

No publication is complete in all respect, and there is always unlimited scope for improvement, in that way this publication is not an exception. We would welcome any suggestion/criticism about this book, which will serve to improve the contents. Nevertheless, we are sure that this book will be extremely useful to students, teachers, vegetable breeders, extension workers and farmers of this country.

Authors

Contents

Chapter 1

Vegetable Research and Development in India: Introduction and History

Vegetables are rich sources of vitamins, minerals, and antioxidants and play an important role in crop diversification, nutritional and economic security. About 40 vegetables are grown commercially in different parts of the country. The production of vegetables in India has increased from 63.8 mt in 1993 to 116 mt in 2006-07. Accordingly, the average productivity has increased from 10.5 t/ha in the year 1992 to 15.47 t/ha. In India, population is increasing at an annual exponential growth rate of 1.93% and to meet the vegetable requirement of 300g per caput per day, vegetable demand by the year 2011 would be ~130 million tonnes and in the year 2021, it would be ~160 million tones to feed the burgeoning population (Fig. 1). Vegetable production in India is distributed across the country, with 56% area in the states like West Bengal, Uttar Pradesh, Orissa and Bihar. The vegetables are grown in ~7% of total cultivated area, whereas, potato, tapioca and sweet potato is grown in about 3.7% of the cropped area. Out of the total vegetable production, potato contributes 28.8%, followed by brinjal (8.6%) and tomato (8.5%). The intensive research and development, and the entry of corporate sector in vegetable seed business have contributed a lot to the vegetable production of the country, but still there exists a vast gap between the national average and potential productivity. There is an ample scope to double the productivity in the next 5-6 years through adoption of improved varieties/hybrids, production and protection technologies. Accordingly, there is a need to achieve the productivity level of 25 t/ha in phased manner with the use of improved technologies including replacement of seeds with improved varieties/hybrids, optimizing input use efficiency, integrated plant nutrient and pest management. The post harvest management and processing of vegetables also needs to be addressed to minimize post harvest losses and improve the availability of vegetables during off-season.

Vegetable research in India

Research on vegetable crops was started by ICAR with the sanctioning of a 'Plant Introduction Scheme' at the IARI, New Delhi in 1948. Later, the Government of India established a Vegetable Breeding Station at Katrain in Kullu Valley, Himachal Pradesh in 1949. This station was transferred to Indian Agricultural Research Institute in 1955 to undertake research on temperate vegetable crops, their seed production technology and to produce seeds of improved varieties.

In the year 1956, a Division of Horticulture was created at the Indian Agricultural Research Institute, New Delhi. Vegetable research was given a priority with the establishment of the Indian Institute

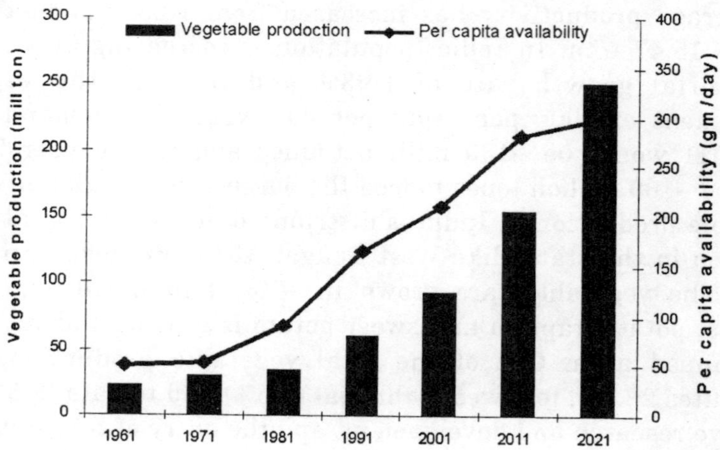

Figure 1. Projected production and availability of vegetables in India

of Horticultural Research (IIHR) at Bangalore with a full-fledged Division of Vegetable Crops in 1968. The vegetable research in India received a further boost when a separate Division of Vegetable Crops and Floriculture was started at IARI, New Delhi in 1970. Establishment of All India Coordinated Vegetable Improvement Project in 1972 gave a further fillip to vegetable research in India. The floriculture research was separated from the Division of Vegetable Crop and Floriculture, IARI, New Delhi in 1982 and the division was then named as Division of Vegetable Crops, which was assigned the research work exclusively on different vegetable crops. In the year 1992, a Project Directorate of Vegetable Research (PDVR) was established at IARI, New Delhi, which was shifted to Varanasi. IIHR, Bangalore started Horticultural Experiment Stations, one each at Ranchi (Bihar) and Godhra (Gujarat) to intensify research on horticultural crops in important tribal areas of the country. Considering the role of vegetables in agriculture, the PDVR was upgraded to Indian Institute of Vegetable Research (IIVR) in 2001. Establishment of department of vegetable crops in various agricultural universities provided further boost in vegetable research and education in India.

Since the start of vegetable research in the country, efforts were made to establish research infrastructure to meet the requirements of vegetable research. Presently, the research on different vegetables is being carried out by central institutes, department of vegetable crops at state agricultural universities, coordinated projects, national research centres, NHRDF, etc.

Central institutes

In the ICAR, vegetable research is being carried out by the Indian Institute of Vegetable Research (IIVR), Varanasi, Indian Agricultural Research Institute (IARI), New Delhi and its Regional Station at Katrain (HP). In addition, strong interdisciplinary research is also underway at the Indian Institute of Horticultural Research (IIHR), Bangalore. Vegetable improvement is also a thrust at Vivekanand Parvatiya Krishi Anusandhan Sansthan (VPKAS), Almora. ICAR Research Complex for North-Eastern Hill Region, Shillong is having a good repository of selected vegetable crops.

National Bureau of Plant Genetic Resources (NBPGR), New Delhi started in 1976 is doing intensive work on exploration, collection, evaluation, conservation and exchange of germplasm of vegetable crops. A National Research Centre on Onion and Garlic (NRCOG) has been established at Pune, which is looking after onion and garlic improvement programmes. In addition, National Horticultural Research and Development Foundation (NHRDF) with its centers at Nasik and Karnal is also carrying research on onion and garlic.

Agricultural universities

Establishment of State Agricultural Universities and starting separate departments of horticulture/vegetable crops bridge the gaps in vegetable research and tackle problems of regional importance. Presently, a total of 43 agricultural universities existing in the country have programmes on research and development on vegetable crops. The Govind Ballabh Pant University of Agriculture and Technology (GBPUA&T) established in 1960 at Pantnagar in Uttaranchal was the first university, which started a Department of Horticulture for varietal improvement and production technology of vegetable crops. Rajasthan Agricultural University (RAU), Bikaner was established in 1962; the University contributes to vegetable research at its Regional Agricultural Research Station (RARS), Durgapura. Orissa University of Agriculture and Technology (OUAT) was also started in 1962 with a department of horticulture including research and teaching on vegetable crops. Punjab Agricultural University (PAU), Ludhiana was established in 1963 has a separate Department of Vegetable Crops, Landscaping and Floriculture, and has played a very significant role in vegetable research. Jawahar Lal Nehru Krishi Vishwavidyalaya (JNKVV), Jabalpur established in 1964 started a separate department of vegetable crops. Aacharya N.G. Ranga Agricultural University (ANGRAU), Hyderabad and University of Agricultural Sciences (UAS), Bangalore were started in 1965 with a department of horticulture.

In the next phase (1969-1972), ten universities, *viz.*, Mahatma Phule Krishi Vidyapeeth (MPKV), Rahuri (1969), Punjabrao Deshmukh Krishi Vidyapeeth (PDKV), Akola (1969), Assam Agricultural University (AAU), Jorhat (1969), Haryana Agricultural University (HAU), Hisar (1970), Tamil Nadu Agricultural University (TNAU), Coimbatore (1971), Rajendra Agricultural University (RAU), Pusa (1971), Marathwada Agricultural University (MAV), Parbhani (1972), Konkan Krishi Vidyapeeth (KKV), Dapoli (1972), Kerala Agricultural University (KAU), Vellanikkara (1972), and Gujarat Agricultural University

(GAU), Anand (1972) were established. Among these, HAU, Hisar; TNAU, Coimbatore; RAU, Pusa; and KAU, Vellanikkara have separate department of vegetable crops. Rests of the universities are undertaking vegetable research and development programmes in the department of horticulture.

In the third phase (1974-1978), four state agricultural universities *viz.*, Bidhan Chandra Krishi Vishwa Vidyalaya (BCKV), Kalyani (1974), Chandra Shekhar Azad University of Agriculture and Technology (CSAUA&T), Kanpur (1975), Narendra Dev University of Agriculture and Technology (NDUA&T), Faizabad (1975), Himachal Pradesh Krishi Vishwa Vidyalaya (HPKV), Palampur (1978) were established. Among these, NDUA&T and HPKV are having separate department for teaching and research on vegetable crops.

Later, five agricultural universities were started after 1980, Birsa Agricultural University (BAU) Ranchi (1982), Sher-e-Kashmir University of Agricultural Sciences and Technology (SKUAST), Srinagar (1982), Dr. YS Parmar University of Horticulture and Forestry (YSPUHF), Solan (1984), University of Agricultural Sciences (UAS), Dharwad (1986), and Indira Gandhi Krishi Vishwa Vidyalaya (IGKV), Raipur (1987). Among these, YSPUHF, Solan and Andhra Horticultural University, Andhra Pradesh are entirely devoted to horticulture and forestry research, whereas, rest of them are having separate department for research and education in the field of vegetable science. Besides, Central Agricultural University opened with head quarter at Nagaland and it has separate college in different region of NEH.

All India Coordinated Research Project on Vegetable Crops (AICRP-VC)

The All India Coordinated Research Project (AICRP) of the ICAR was conceptualized during the reorganization of the council in 1955, to make it an effective organ of agricultural research in the country. The AICRPs provided a major and important forum for placing agricultural research in the country on a sound basis. It operates on national basis and multi-location research approach to provide scope and opportunities for wider evaluation of the proven results. Being a multi-location testing mechanism, AICRPs are able to cover a wide range of agro-ecological conditions. The centers of research under these projects are located both in the agricultural universities and the research institutes. This provides a functional linkage between the institutions and the scientists working on related problems. The objectives of the AICRPs are to undertake problem oriented applied research and testing of the technology under different agroclimatic conditions with a view to solve the national problems in the shortest possible time.

Based on these principles, an All India Coordinated Research Project on Vegetable Crops (AICRP-VC) was started by ICAR in 1970-71, during the Fourth Five-Year Plan. Headquarter of the project was at the division of vegetable crops, IARI, New Delhi, which was shifted to IIVR, Varanasi. The project was having 1 co-ordinating centre, 7 main centers, 18 sub-centres, and 26 voluntary centers. In the XIth five year plan (2007-2012), due to the merger of NSP-VC with AICRP-VC, the total number of centres (excluding the voluntary centres) under AICRP-VC is now 29; RAU, Pusa, CAU, Passighat and IARI-RS, Karnal are the three new centres, which have been added to AICRP-VC.

Coordinating Centre

Indian Institute of Vegetable Research, Varanasi,

Main centers

1. Indian Agricultural Research Institute, New Delhi
2. IARI Regional Station, Katrain, Kullu
3. Indian Institute of Horticultural Research, Bangalore
4. Punjab Agricultural University, Ludhiana
5. Rajendra Agricultural University, Sabour
6. Mahatma Phule Krishi Vidyapeeth, Rahuri
7. Tamil Nadu Agricultural University, Coimbatore

Sub-centers

1) G.B. Pant University of Agriculture and Technology, Pantnagar
2) C.S. Azad University of Agriculture and Technology, Kanpur
3) Assam Agricultural University, Jorhat
4) Orissa University of Agriculture and Technology, Bhubaneswar
5) Jawaharlal Nehru Krishi Vishwa Vidyalaya, Jabalpur
6) S.K. University of Agriculture and Technology, Srinagar
7) Himachal Pradesh Agricultural University, Palampur
8) Andhra Pradesh Agricultural University, Lam, Guntur

9) Udaipur University, Durgapura, Jaipur
10) Bidhan Chandra Krishi Vishwavidyalaya, Kalyani
11) Marathwarda Krishi Vishwavidyalaya, Ambajogai
12) Haryana Agricultural University, Hisar
13) Narendra Dev University of Agriculture and Technology, Faizabad
14) Kerala Agricultural University, Vellanikkara
15) Andhra Pradesh Agricultural University, Hyderabad
16) Gujarat Agricultural University, Junagadh
17) University of Agricultural Science, Dharwad
18) Indira Gandhi Krishi Vishwavidyalaya, Raipur

In addition, 26 centers, involving central research institutes, state agricultural and traditional universities and some public and private research organizations are participating in the cooperative research programme of the project on voluntary basis.

The first major task of the AICRP-VC was multi-location testing of varieties developed by different research institutes and agricultural universities at the centres located in different agroclimatic conditions to identify varieties suitable for different areas. Later, research was undertaken on agro-techniques, like plant population density, fertilizer requirements and chemical weed control. Research programmes were also undertaken on chemical control of diseases and insect pests. Improved varieties of crops are the backbone for increasing productivity and production. Presently, a total of 23 vegetable crops are covered under AICRP-VC. However, the major emphasis is laid on crops like tomato, brinjal, chillies, cauliflower, cabbage, pea, onion, muskmelon, watermelon, okra, French bean, pumpkin, carrot, radish, garlic, etc. This intensive programme has led to the development of 366 varieties/hybrids, 105 production and 95 protection technologies in 23 vegetable crops. IIVR alone has developed 43 varieties/hybrids in 12 vegetable crops with 100 production and protection technologies. Workshops/group meetings are also organized to identify varieties and improved package of practices. Till date, a total 17 workshops/ group meetings have been organized. In the subsequent phase of development, several new centres were included under this project. The basic aim behind these new inclusions is to cover broadly classified eight agro-climatic zones of India (Table 1) under the AICRP trials.

Table 1 : Agro-climatic zones of India and states under each zone.

Zone	Geographical region	State under the zone
I	Humid Western Himalayan Region	Jammu & Kashmir (J&K), Himachal Pradesh and Uttaranchal
II	Humid Bengal-Assam Basin	West Bengal and Assam
III	Humid Eastern Himalayan and Bay Islands	Sikkim, Meghalaya, Manipur, Nagaland, Mizoram, Tripura, Arunachal Pradesh and Andaman & Nicobar Islands
IV	Sub-Humid Sutlej Ganga Alluvial Plain	Punjab, U.P. and Bihar
V	Sub-Humid to Humid Eastern and South Eastern Uplands	Chattisgarh, Orissa and Andhra Pradesh
VI	Arid Western Plain	Rajasthan, Gujarat, Haryana and Delhi
VII	Semi Arid Lava Plateau and Central High Lands	Madhya Pradesh and Maharashtra
VIII	Humid to Semi Arid Western Ghats and Karnataka Plateau	Karnataka, Tamil Nadu and Kerala

Private organizations

In India, most of the vegetable research is being carried out by public institutions. However, from 1969 onwards, there has been an effort to start R&D activities by some private companies with foreign collaboration. Some private companies conducting research include Indo-American Hybrid Seed Company, (Bangalore), Mahyco, (Jalna), Nath Seeds, (Aurangabad), Bejo-Sheetal Seeds, (Jalna), Namdhari Seeds, (Bangalore), Novartis India Limited, Pune. A number of open pollinated varieties and hybrids in vegetable crops have been developed by different private seed companies which have performed well under multi-location testing under AICRP-VC.

National Seed Project on Vegetables (NSP-Vegetables)

Research back up in terms of development and availability of information on appropriate seed production method, techniques of seed quality maintenance, storage and seed health care is most vital and essential to effectively execute the breeder, foundation and certified seed production programmes. Realizing the importance of seed in increasing vegetable production, the Government of India launched a National Seed Project (NSP) in October 1976 with World Bank assistance. A separate NSP on vegetables (NSP-VC) was approved and initiated by PDVR, Varanasi in 1994. Under NSP-VC, 14 centres are working in different parts of the country. This includes 10 centres in the state agricultural universities and four under ICAR system. These 14 breeder seed production units at Varanasi, Karnal, Katrain, Banglore, Ludhiana, Kanpur, Rahuri, Solan, Coimbatore, Jabalpur, Lam, Pantnagar, Bhubaneswar and Raipur dealing with vegetables (excluding potato) are being monitored by IIVR, Varanasi. The production of breeder seed of different varieties of vegetable crops of last 5 years against national indent is approximately 20-40% higher than the required (Annexure I). In addition to the responsibility of production of nucleus and breeder seed of varieties and parental lines of hybrids, the production programmes are duly supported by nucleus seed production and research on various aspects of seed technology like isolation requirements, genetic purity tests, seed physiology and seed health care, processing, packaging, storage, etc. However, in the XIth five year plan (2007-2012), National Seeds Project on Vegetable Crops (NSP-VC) has been merged with AICRP-VC, in order to effectively utilize the man power, since the work of NSP-VC involved the production of vegetable seeds as well as research on seed related aspects.

Conclusion

Although commendable progress has been made on the research front and a number of technologies have been developed, poor adoption of such technologies has always been a major handicap in increasing productivity. Therefore, promotion of developed technologies through various mechanisms is a key to harness the available technologies, thus ensuring the increased productivity and profitability to the farmers. Proactive government policies for development of infrastructure and law enforcement, which support execution of promotional activities, are equally pertinent to achieve higher productivity of vegetables.

Chapter 2

Solanaceous Vegetables

TOMATO

Solanum lycopersicum
Family: Solanaceae
Chromosome no.: 2n=24

Tomato is one of the most popular vegetables grown all over the world. It belongs to family Solanaceae, genus *Solanum* and species *lycopersicum*. It is considered to be a native of Peru (South America), and is believed to spread to North America primarily by migrating birds. The wild forms of tomato are primarily found in Mexico and the name tomato itself comes from the Aztec word Tomatl. In India, with its wider coverage in comparison to other vegetables, Uttar Pradesh, Karnataka, Maharashtra, Haryana, Punjab and Bihar are the leading tomato growing states. It is grown as annual plant and small and marginal farmers are getting good income through tomato cultivation. Tomato is generally used as salad, culinary vegetable and for processed products. Several processed products like paste, puree, syrup, juice, ketchup, drinks, whole peeled tomato etc. are prepared on large-scale. On the basis of growth habit, tomato plants can be grouped as determinate (growth restricted due to conversion of terminal part as flower) or indeterminate (growth is continued). Although, tomato is a warm season crop with 21-23°C as optimum temperature for its vegetative growth, but plants can not withstand frost. The fluctuation of night temperature and light intensity largely affect fruit set, lycopene content and nutritive value of the fruits. Tomato can be at its best in sandy loam soil with rich in organic matter and soil pH 6.0-7.0.

The key characteristics of the flowers are stamen usually 5, opening from base to top. Dehiscence of anther occurs 1-2 days after the opening of corolla. Tomato is a typical day-neutral plant. It is mainly self-pollinated but certain amount of cross-pollination also occurs depending on the variety and environment, especially, when the stigma protrudes outside the level of anther (exerted). The optimum range of night temperature for fruit set is 15-20°C.

In northern India, seeds are sown in nursery bed in the month of June-July (Autumn-Winter season) or November (summer crop), but in southern parts of the country sowing is done throughout the year. In hills, suitable time of sowing is March-April. To raise the seedling for one hectare, 225-250 m^2 area is required and 400-500 g of open pollinated and 175-200 g seeds of hybrid is needed. Seedlings become ready for transplantation in 4-5 weeks after seed sowing. Fruits can be harvested 80-100 days after planting depending upon the season. Harvesting of fruits mainly depend on marketing/processing purposes. Red ripe tomatoes are preferred for local markets and processing units, while fully matured green, red or pink are preferred for distant markets and storage. Most destructive diseases of tomato are damping off (nursery disease), TLCV, bacterial wilt, *Fusarium* wilt and blight.

Most of the tomato varieties have been developed through introduction, pure line selection and pedigree selection. Hybrid seeds are largely produced by hand emasculation and hand pollination. For this purpose, both the parents are grown in the ratio of 12 female line:1 male line. The cost of hybrid seeds are generally minimized by a use of the male sterile line.

Varieties

A large number of tomato varieties have been evolved by research organizations and different universities. These varieties have desirable attributes like early fruiting, high yield, better fruit size, colour and disease/pest resistance. The details of the varieties are given as under (for photographs see plate 1).

SIOUX

This variety is derived through selection from an American introduction and recommended by IARI, Regional Station, Katrain, and notified by CVRC Notification nos. 716, 361(E), and 440 dated 20.2.1970, 30.6.1973 and 21.08.1975, respectively. Sioux is an indeterminate variety. Fruits are uniform red, round, smooth, small and less-seeded. This variety has been recommended for cultivation throughout the country for summer and spring season and has yield potential of 195-220 q/ha.

PUSA RUBY

This variety has been developed at IARI, New Delhi and notified by CVRC Notification nos. 716, 361(E), 440, and 540(E) dated 20.2.1970, 30.6.1973, 21.08.1975, and 24.07.1985, respectively. This is an early maturing variety derived through selection from a cross Sioux × Improved Meeruthi. Plants are indeterminate with a height of 80-85 cm and spreading habit with few hardy branches. Fruits are flat to round, slightly lobed (4-5 locules), medium-sized, uniform red at maturity with slightly acidic pulp, suitable for fresh consumption and processing. It is cultivated in autumn, winter and spring-summer seasons in north India. This variety has been recommended for cultivation throughout the country and has yield potential of 280-300 q/ha.

S-12

This variety has been developed at PAU, Ludhiana and notified by CVRC Notification no. 716, and 1004 dated 22.02.1970, and 23.03.1978, respectively. Plants are dwarf with vigourous growth. Fruits are medium-sized and uniform red. This variety has yield potential of 450-480 q/ha.

MARGLOBE

This variety is an introduction, recommended by IARI, New Delhi and notified by CVRC Notification nos. 716, 361(E), 440, and 13 dated 20.2.1970, 30.6.1973, 21.08.1975, and 19.12.1978, respectively. It is a mid-season variety. Plants are 70-80 cm tall, vigorous, erect and indeterminate in growth habit. Leaves are large and broad with dense foliage. It has typically 4 or 5 flowers on truss, which bears 3 or 4 fruits. Fruits are nearly round, deep scarlet on ripening, 3-5 locules, juicy, mild acidic, thick-walled, smooth, solid and less-seeded. Fruits have dark green shoulder. Diameter of ripe fruit is about 7.5 cm and weight 100-150 g. This variety has yield potential of 200-250 q/ha.

SL-120

This variety has been developed at IARI, New Delhi and notified by CVRC Notification no. 716 dated 20.02.1970. Plants are semi-determinate, spreading, late maturing with dark green foliage. Fruits are flatttish-round, attractive, medium-large, uniform red, less-acidic, less-seeded, resistant to nematode and suitable for winter and summer season cultivation. This variety has been recommended for general cultivation throughout the country and has yield potential of 300-320 q/ha.

BEST OF ALL

This variety is an introduction and recommended by IARI, New Delhi, and notified by CVRC Notification nos. 361 (E), 440, and 13 dated 20.2.1970, 30.6.1973, and 21.08.1975, respectively. It is an indeterminate and mid-season variety. Plant are erect and height ranges from 100 to 200 cm. Fruits are medium in size, scarlet red, round, juicy, small scar, thin skinned, 3 loculed with green shoulder and borne in clusters of 4 to 5 fruits. This variety has yield potential of 300-315 q/ha.

PUSA EARLY DWARF

This variety has been developed at IARI, New Delhi and notified by CVRC Notification nos. 361(E), 440, and 540(E) dated 30.6.1973, 21.08.1975, and 24.07.1985. This variety has been developed through selection from a cross Improved Meeruthi × Red Cloud. Plants are determinate, early maturing and dwarf with compact fruiting. Fruits are flatish-oblate, smooth, medium-large, red, furrow with 5-6 locules and suitable for rainy season cultivation in plains and March sowing in hills. This variety has been recommended for cultivation in Delhi and surrounding areas and has yield potential of 300-325 q/ha.

HS-101

This variety has been developed through hybridization (Sel-2-3 × Exotic cultivar) followed by pedigree selection at HAU, Hisar and notified by CVRC Notification nos. 786, and 1004 dated 02.02.1976, and 23.03.1978. Plants are determinate, dwarf (55 cm) with dark green foliage and bear round red fruits. This variety has been recommended for cultivation throughout the country and has yield potential of 310-325 q/ha.

SOLAN GOLA

This variety has been developed at YSPUHF, Solan and notified by CVRC Notification no. 440 dated 21.08.1975. Plant are indeterminate having round, firm fruits of medium size. It has yield potential of 375-390 q/ha.

HS-102

This variety has been developed at HAU, Hisar and notified by CVRC Notification no. 786 dated 02.02.1976. It is suitable for both winter and summer seasons and adapted to Haryana state. Plants are 45-50 cm tall and determinate in growth habit. Its fruits are borne in cluster of 3-4, small to medium in size, and 3-4 loculed and uniform red on ripening. This variety bears 40-50 fruits/plant and has yield potential of 250-270 q/ha.

CO-2

This variety has been developed through pure line selection (USSR line) at TNAU, Coimbatore and notified by CVRC Notification no. 13 dated 19.12.1978. Plants are 70 cm tall, semi dwarf, erect in growth habit with moderate branching, well-spread and needs no staking. Stem is light green and foliage is dark green. Flowering occurs 65-70 days and first picking starts 100 days after transplanting. Fruits are ovate, medium in size, smooth, green when unripe and bright red on ripening. Fruits are devoid of cracking. It bears 4-5 fruits/truss and 20-25-fruits/plant with an yield potential of 410-425 q/ha in 6-8 harvesting.

KECKRUTH

This introduced variety has been developed through hybridization (Kecki Methi × Ruthger) followed by pedigree selection and notified by CVRC Notification nos. 13, and 563 (E) dated 19.12.1978, and 30.08.1991. It is 100 cm tall and stem and foliage are green. Flowering takes place 93 days after transplanting. Its fruits are round, medium to large (70-80 gm), deep red on maturity with 3-4 locules and 0.3 cm thick pericarp. It has yield potential of 200-250 q/ha.

KECHRUTH AGETI

This variety is an introduction and notified by CVRC Notification nos. 13, and 563 (E) dated 19.12.1978, and 30.08.1991. It is a derivative of a cross Keckruth × Ruthger. Plants are 100 cm tall with deep green foliage. Flowering starts 85 days after transplanting. Fruits are round, medium in size (40-50 gm) with green shoulder, which disappears on ripening with 3-4 locules and 0.24 cm thick pericarp. It has yield potential of 225-250 q/ha.

PUNJAB CHHUHARA

This variety has been developed through a cross between an introduced line from Israel and Shimla Gold at PAU, Ludhiana and notified by CVRC Notification no. 13 dated 19.12.1978. Plants are determinate, dwarf and bushy with dense luxuriant foliage. Fruits are red, medium, pear-shaped, firm, fleshy, bilocular, less-acidic with thick pericarp and less-seeded and suitable for long distance transportation. Plants are highly susceptible to TLCV. This variety has been recommended for cultivation in the areas where TLCV is not a major problem and has yield potential of 230-250 q/ha.

PUNJAB TROPIC

This variety is an introduction from USA, recommended by PAU, Ludhiana and notified by CVRC Notification no. 13 dated 19.12.1978. Plants are 100 cm tall, indeterminate growth habit with luxuriant growth. Stem and foliage are dark green. Flowering starts 95 days after sowing. Fruits are large (100 g), round, red on maturity, borne in clusters (4-6 fruits) and remain totally covered with foliage. It has late fruit setting with slow development of fruits. It has yield potential of 225-240 q/ha in 145 days of crop duration.

NARENDRA TOMATO-1 (NDT-5)

This variety has been developed at NDUA&T, Faizabad and notified by CVRC Notification no. 115(E) dated 10.02.1996. Plants are indeterminate with hairy leaves. Fruits are round, medium to large size (85-110 g), smooth, juicy. First fruit can be harvested in 75-80 days. It has yield potential of 400-450 q/ha.

KALIANPUR ANGOORLATA

This variety is a selection from local germplasm, developed at CSAUA&T, Kanpur and notified by CVRC Notification no. 2(E) dated 03.01.1983. Plants are of creeping in nature, 200-300 cm tall with indeterminate growth habit. Leaves are medium, light green with spaced leaflets. Fruits are small and oblong with 8-10 fruits/truss, firm and thick skin, less juicy and 8-10 days keeping quality at room temperature. Immature fruits are yellowish green and red on ripening. Plants are moderately resistant to leaf curl, and resistant to root knot nematode. It has yield potential of 480-500 q/ha.

JUNAGADH RUBY

This variety has been developed through hybridization (Big Silari × Pusa Ruby) followed by pedigree selction at GAU, Junagadh and notified by CVRC Notification no. 596(E) dated 13.08.1984. Plants are semi-spreading in growth habit with purple attractive foliage. Leaves are medium in size and dark green. Fruits are born in cluster of 4-6 fruits/truss, round without grooves, 4 loculed, orange red on ripening with pinkish red flesh. First picking is done at 85-90 days. On an average, it gives 20 pickings at short intervals. It has yield potential of 280-300 q/ha.

MARUTHAM (CO-3)

This variety developed at TNAU, Coimbatore and notified by CVRC Notification no. 596(E) dated 13.08.1984 has been derived by EMS induced mutation in CO-1 variety. Plants are determinate, erect, compact, dwarf and bear 30-40 fruits. Fruits are round globular, medium (45-50 g), smooth, parrot green (unripe) and borne in cluster with 3.2% T.S.S. First picking starts 80-85 days after transplanting. This variety has been recommended for cultivation in J&K, H.P., Uttaranchal, Punjab, U.P., Bihar, Chattisgarh, Orissa, A.P., Rajasthan, Gujarat, Haryana, Delhi, M.P., and Maharashtra and has yield potential of 450-470 q/ha in 100-105 days of crop duration.

AZAD T-2 (KS-2)

This variety has been developed through hybridization (HB-5 × Kuber) followed by pedigree selection at CSAUA&T, Kanpur and notified by CVRC Notification no. 295(E) dated 09.04.1985. Plants are determinate, compact with leaves of curved (inwards) margin. Fruits are small, round with 3-4 locules, slightly furrowed green at stem end, red flesh, 5.66% TSS. Per 100 g of fruits contain 3.68 mg lycopene and 15 mg vitamin C. First picking starts 65-70 days after transplanting. Plants are moderately resistant to leaf curl virus with 75-80 days of active fruiting period. This variety has been recommended for cultivation in U.P. and has yield potential of 330-350 q/ha in 130-135 days of crop duration.

PANT BAHAR (AC-238)

This variety has been developed through selection from a germplasm line (AC-238) at GBPUA&T, Pantnagar and notified by CVRC Notification no. 540 (E) dated 24.07.1985. Plants are indeterminate, bushy (90 cm), profusely branched with light green foliage. Fruits are flattish-round, medium with 5-6 locules, slightly ridged with good storage and processing quality. First picking starts 75-80 days after transplanting. This variety has been recommended for cultivation in U.P. and has yield potential of 250-270 q/ha.

PUNJAB NR-7

This variety has been developed from the segregating population of a cross (S-12 × NMR-1) at PAU, Ludhiana and notified by CVRC Notification no. 165 (E) dated 06.03.1987. It is a determinate variety, has bold, round flat fruits with an excellent blend of TSS and acidity. This variety requires less water and is fairly tolerant to sub-optimal temperatures. It has been recommended for cultivation in nematode infested soils and has yield potential of 380-400 q/ha.

PUSA GAURAV

This variety has been developed from segregating generation of a cross involving exotic lines (Galmour × Watch) at IARI, New Delhi and notified by CVRC Notification no. 10(E) dated 01.01.1988. Plants are dwarf, bushy with moderate foliage cover. Fruits are smooth, elliptical (egg-shaped) and borne in clusters. The unripe fruits are firm with thick flesh (0.6 cm) with two well-filled locules facilitating easy transportation over long distances. Whole fruit is suitable for processing and canning as it is without neck constriction and has higher TSS (6%) and better keeping quality at room temperature. It has yield potential of 330-350 q/ha.

PANT T-3

This variety has been developed through pure line selection at GBPUA&T, Pantnagar and notified by CVRC Notification no. 1135(E) dated 01.12.1988. Plants are semi-determinate with thick round hairy stem and dark green foliage. Fruits are round, smooth and uniform red, suitable for processing. It is suitable for winter season cultivation, first picking starts 75-80 days after transplanting. This variety has been recommended for cultivation in J&K, H.P., Uttaranchal, West Bengal, Assam, Punjab, U.P., Bihar, Chattisgarh, Orissa, A.P., Rajasthan, Gujarat, Haryana, Delhi, M.P., and Maharashtra and has yield potential of 325-350 q/ha.

VASUNDHARA (ATV-1)

This variety has been developed at MPKV, Rahuri and notified by CVRC Notification no. 915(E) dated 06.11.1989. Fruits are bright red and each weighing 90-100 g. This variety has yield potential of 450-470 q/ha.

ATH-1

This hybrid has been developed at MPKV, Rahuri and notified by CVRC Notification no. 915(E) dated 06.11.1989. Fruits are red and medium-sized and each fruit weighing 110 g. It has yield potential of 450-470 q/ha.

ARKA VIKAS (SEL-22)

This variety has been developed through selection from an American introduction (Tip Top) at IIHR, Bangalore and notified by CVRC Notification no. 386(E) dated 15.05.1999. Plants are semi-determinate with narrow dark green foliage and good canopy. Fruits are medium-large (80-90 g), uniform deep red and oblate with light green shoulder, suitable for fresh market and tolerant to mosaic. It is well adapted to both rainfed and irrigated conditions. This variety has been recommended for cultivation in J&K, H.P., Uttaranchal, West Bengal, Assam, Punjab, U.P., Bihar, Chattisgarh, Orissa, A.P., Rajasthan, Gujarat, Haryana, Delhi, M.P., Maharashtra, Karnataka, Tamil Nadu, and Kerala and has yield potential of 400-425 q/ha in 140 days of crop duration.

UTKAL PALLAVI

This variety has been developed at OUA&T, Bhubaneshwar and notified by CVRC Notification no. 860(E) dated 25.11.1992. Plants are determinate with curly leaf. Fruits are borne in cluster of 6-12, medium in size, pear-shaped, nipple at blossom end, firm and fleshy. It is resistant to bacterial wilt and nematode. It also perform well in saline soil. It matures 85 to 90 days after sowing and has yield potential of 375-380 q/ha.

HISAR LALIT (NT-8)

This variety has been developed by hybridization (HS-101 × Resistant Bangalore) followed by pedigree selection at HAU, Hisar and notified by CVRC Notification no. 617(E) dated 17.08.1993. Plants are semi-determinate with early fruiting. Fruits are round and medium to large. It is resistant to root knot nematode and recommended for cultivation in root knot nematode infested areas. It has yield potential of 300-325 q/ha.

HISAR ARUN (SEL-7)

This variety has been developed through modified pedigree method from a cross (Pusa Early Dwarf × K-1) at HAU, Hisar and notified by CVRC Notification no. 636(E) dated 02.09.1994. It is an early maturing, determinate, dwarf, erect, with cut leave and synchronized clustered flowering bearing 15-20 fruits. Fruits are round, red, medium, average weight 65-70 g, 4-6 locules with deep red flesh. First picking starts 60-65 days after transplanting. This variety has been recommended for cultivation in J&K, H.P., Uttaranchal, West Bengal, Assam, Punjab, U.P., Bihar, Chattisgarh, Orissa, A.P., Rajasthan, Gujarat, Haryana, Delhi, M.P., Maharashtra, Karnataka, Tamil Nadu, and Kerala and has yield potential of 170 q/ha and 280 q/ha early and total yield, respectively, in 80-85 days of crop duration.

NARENDRA TOMATO-2 (NDT-120)

This variety has been developed at NDUA&T, Faizabad and notified by CVRC Notification no. 115(E) dated 10.02.1996. Plants are determinate with curly leaves. Fruits are round, pointed in back side, medium to large size, smooth and solid with good keeping quality and single fruit weight of 60-75 g. First fruit can be harvested in 75-80 days. It has yield potential of 400-450 q/ha.

PUSA HYBRID-2

This hybrid has been developed through heterosis breeding at IARI, New Delhi and notified by CVRC Notification no. 115(E) dated 10.02.1996. Plants are compact, semi-determinate with good foliage cover having prolific bearing. Fruits are round to flatish round, firm, smooth and attractive with uniform red at maturity. Plants are highly tolerant to root knot nematode. It has been recommended for cultivation in humid western Himalayan regions, sub-humid Sutluj-Ganga alluvial plains and Western plains, semi-arid plateau and has yield potential of 600-625 q/ha.

SOLAN SHAGUN

This hybrid has been developed at YSPUH&F, Solan and notified by CVRC Notification no. 115(E) dated 10.02.1996. It is semi-indeterminate, erect with deep green foliage. First picking can be taken at 70 days. Fruits are slightly oval and deep red, average fruit weight 65 g and suitable for long distance transportation. Plants are tolerant to Alternaria blight and buckeye rot. It has yield potential of 490-500 q/ha.

PUSA DIVYA (KT-4)

This hybrid has been developed at IARI, New Delhi and notified by CVRC Notification no. 98(E) dated 08.02.1997. It has been developed by utilizing closed anther mutant with potato leaf marker gene (functional male sterility) as female parent. Plants are indeterminate with profuse branching. Fruits are firm, thick-skinned, round to oval, green with dark green stem end and turn dark red on ripening. Due to thick skin, fruits have longer shelf-life with suitability for long distance transportation. Maturity period is 70-90 days from transplanting and 135-150 days from seed sowing in hills. It has been recommended for cultivation in Uttar Pradesh, Punjab and Bihar and has yield potential of 350-450 q/ha.

PUSA HYBRID-4 (DTH-4)

This hybrid has been developed at IARI, New Delhi and notified by CVRC Notification no. 98(E) dated 08.02.1997. Plants are determinate, compact with good dark green foliage. Fruits are attractive, round, smooth, average fruit weighing 70-80 g with three locules, thick pericarp and uniform ripening. It has field resistance to root-knot nematode and suitable for semi-Arid plateau and Central High Lands. It has yield potential of 425-450 q/ha.

UTKAL KUMARI (BT-10)

This variety has been developed by hybridization (Selection-22 × BT-2) followed by pedigree selection at OUA&T, Bhubaneshwar and notified by CVRC Notification no. 843(E) dated 21.09.1998. It is a bacterial wilt resistant and indeterminate variety. Its fruits are round, red, medium-large with 5-6 days of shelf-life. First picking starts 90-95 days after sowing. It has been recommended for cultivation in Chhattisgarh, Orissa, A.P., Rajasthan, Gujrat, Haryana and Delhi and has yield potential of 400-415 q/ha in 90 to 100 days (after transplanting) of crop duration.

SOLAN VAJR

This variety has been developed at YSPUH&F, Solan and notified by CVRC Notification no. 843(E) dated 21.09.1998. Plants are indeterminate with medium sized, firm and heart-shaped fruits. This variety has yield potential of 450-475 q/ha.

MUKTHI (LE-79-5)

This variety has been developed through pure line selection at KAU, Vellanikkara and notified by CVRC Notification no. 843(E) dated 21.09.1998. Plants are determinate with pubescent stem and normal leaf. Fruits are round, white and of medium size with average fruit weigh of 58 g. First picking starts 75 days after sowing. It is resistant to bacterial wilt and fruit cracking. It has been recommended for cultivation in J&K, H.P., Uttaranchal, Chhattisgarh, Orissa, A.P., M.P., Maharashtra, Karnataka, Tamil Nadu, and Kerala and gives an avarage yield of 430-450 q/ha in 95-100 days of crop duration.

VASUNDHARA (HYBRID-28)

This hybrid has been developed at MPKV, Rahuri and notified by CVRC vide notification no. 1135(E) dated 15.11.2001. Plants are determinate with Jasper red fruits, each weighing 70-75 g with good storability. It is resistant to virus and blight and has yield potential of 500-525 q/ha.

PARBHANI YASHASHRI (SEL-14)

This variety has been developed at MPKV, Rahuri and notified by CVRC Notification no. 1135(E) dated 15.11.2001. Plants are semi-determinate, suitable for kharif and rabi season, having fruits with good taste. It is tolerant to virus and blight and has yield potential of 600-625 q/ha.

SWARNA BAIBHAV

This variety has been developed through heterosis breeding at HARP, Ranchi and notified by CVRC Notification no. 642 (E) dated 31.05.2004. Fruits are round, deep red, firm, each with 140-150 g weight having high pulp and suitable for long distance transportation and processing. First harvest can be taken at 55-60 days after transplanting. It is resistant to bacterial wilt and has been recommended for August-September sowing in Jharkhand and Bihar and has yield potential of 700-725 q/ha.

KASHI VISHESH (H-86)

This TLCV resistant variety has been developed through backcross pedigree selection using *L. hirusutum* f. *glabratum* B-6013 as a donor parent at IIVR, Varanasi and notified by CVRC Notification no. 597(E) dated 25.04.2006. It has determinate growth habit, dark green foliage, red spherical medium to large fruits each weighing 80 g. First harvest can be taken at 70-75 days after transplanting It has been recommended for cultivation in J&K, H.P., Uttaranchal, Punjab, U.P., Bihar, Jharkhand, Chhatishgarh, Orissa, A.P., Karnataka, Tamil Nadu and Kerala and has yield potential of 430-450 q/ha.

KASHI AMRIT (DVRT-1)

This variety has been developed through backcross pedigree method from a cross (Sel-7 × *L. hirsutum* f. *glabratum* B-6013) at IIVR, Varanasi and notified by CVRC Notification no. 597(E) dated 25.04.2006. It is a determinate variety bearing medium to large size, round attractive red fleshy fruits each weighing ~100 g. It is suitable for cultivation in TLCV prone areas and has yield potential of 620-635 q/ha in 115-120 days of crop duration.

KASHI ANUPAM (DVRT-2)

This variety has been developed through through backcross pedigree method from a cross (Sel-7 × *L. hirsutum* f. *glabratum* B-6013) at IIVR, Varanasi and notified by CVRC Notification no. 597(E) dated 25.04.2006. Plants are determinate bearing attractive red, flattish round, large fruits (slightly indented at blossom end) with 5-6 locules. It has medium maturity (75-80 days after transplanting). It has been recommended for cultivation in Rajasthan, Gujarat, Haryana and Delhi and has yield potential of 570-600 q/ha.

KASHI HEMANT (IIVR SEL-1)

This variety has been developed through pedigree selection from a cross (Sel-18 × Flora Dade) at IIVR, Varanasi and notified by CVRC Notification no. 597(E) dated 25.04.2006. Plants are determinate, fruits attractive red and round-shaped, on an average this variety bears about 30 fruits/plant and each fruit weighing 80 to 85 g. It has been recommended for cultivation in Chattisgarh, Orissa, A.P., M.P., and Maharashtra and has yield potential of 400-425 q/ha.

KASHI SHARAD (IIVR SEL-2)

This variety has been developed through pedigree selection from a cross (MTH-6 × Kalyani Eunish) at IIVR, Varanasi and notified by CVRC Notification no. 597(E) dated 25.04.2006. It has indeterminate growth habit, broad leaves, fruits are attractive red, slightly oval and firm with thick pericarp resulting to longer shelf-life, numbers of fruits/plant vary from 35 to 40 and each fruit weighing 90-95 g. It has been recommended for cultivation in J&K, H.P. and Uttaranchal and has yield potential of 480-500 q/ha.

ARKA MEGHALI

This variety has been developed through pedigree selection from a cross (Arka Vikas × IIHR-554) at IIHR, Bangalore and notified by CVRC Notification no. 2035(E) dated 28.11.2006. Plants are semi-determinate and fruits are superior to variety Pusa Ruby in size (63 g) with flesh thickness and firmness. It is suitable for rainfed conditions and has yield potential of 270-290 q/ha.

SWARNA NAVEEN

This variety has been developed through pure line selection at HARP, Ranchi and notified by CVRC Notification no. 2035(E) dated 28.11.2006. Fruits are deep red, oblong, weighing 60-70 g with 5% TSS. Plants are resistant to bacterial wilt. First harvest can be taken 50-60 days after transplanting. It has been recommended for July-September and February-April sowing in Jharkhand and Bihar and adjoining areas and has yield potential of 580-600 q/ha.

SWARNA LALIMA

This variety has been developed through pure line selection at HARP, Ranchi and notified by CVRC Notification no. 2035(E) dated 28.11.2006. Fruits are deep red and round, weighing 120-125 g with 4% TSS. First harvest can be taken 60-65 days after planting. Plants are resistant to bacterial wilt. It is suitable for July-September and February-April sowing and recommended for cultivation in Jharkhand and Bihar and has yield potential of 500-550 q/ha.

SWARNA SAMPADA

This hybrid has been developed through heterosis breeding at HARP, Ranchi and notified by CVRC Notification no. 2035(E) dated 28.11.2006. Fruit are firm, round and red, weighing 120-130 g with 4.5-5.0% TSS and 0.35-0.40% acidity and contain more pulp. Plants are resistant to bacterial wilt and early blight disease. First harvest can be taken 55-60 days after planting. It has been recommended for August-September sowing in Punjab, Uttaranchal, Jharkhand, Bihar and Uttar Pradesh and has yield potential of 600-650 q/ha.

MANI LEIMA

This variety has been developed at ICAR Research Complex for NEH region, Barapani and notified by CVRC Notification no. 2035(E) dated 28.11.2006. It is an early maturing, determinate and multibranched variety. Fruits are medium size, oval, nipple-shaped and bilocular. Fruits are suitable for processing and long distance transportation. It is resistance to fruit cracking and tolerant to TLCV. It is also tolerant to moisture stress and suitable for spring summer season. It has yield potential of 630-650 q/ha.

AZAD T-6

This variety has been developed at CSAUA&T, Kanpur and notified by CVRC Notification no. 597 (E) dated 25.4.2006. Plants are determinate bearing red, round, uniform, smooth and firm fruits. It has yield potential of 250-275 q/ha.

GUJARAT TOMATO-2 (GT-2)

This variety has been developed through hybridization (Angurlata x Punjab Chhuhara) followed by pedigree selection at AAU, Anand and notified by CVRC Notification no. 597 (E) dated 25.4.2006. It is determinate bearing heart-shaped, medium-sized fruits, which are dark red at maturity and possess good shelf-life. Plants are tolerant to leaf curl virus and leaf blight disease. It has been recommended for cultivation in Gujarat and Saurashtra regions and has yield potential of 330-350 q/ha.

PUNJAB KESRI

Punjab Kesri an early maturing variety has been developed through hybridization (Punjab Tropic × EC-55055) followed by pedigree selection at PAU, Ludhiana and identified for release through AICRP-VC in 1987. Plants are determinate, dwarf (60 cm) with profusely branched thick stem and dense foliage, which protect fruits from sunburn. Fruits are oval-round, dark red and ocular with thick pericarp, average weight of 55 g. It has been recommended for cultivation in J&K, H.P., Uttaranchal, Punjab, U.P., Bihar, Chhattisgarh, Orissa, A.P, Rajasthan, Gujarat, Haryana and Delhi and has yield potential of 560-580 q/ha.

LA BONITA

La Bonita is an American variety recommended by NBPGR, New Delhi and identified for release through AICRP-VC in 1987. It is determinate and bushy. Fruits are smooth, oblong with thick pericarp and 2-3 locules, suitable for long distance transportation. It has been recommended for cultivation in J&K, H.P., Uttaranchal, Punjab, U.P., Bihar, Chhattisgarh, Orissa, A.P., M.P., and Maharashtra and has yield potential of 380-400 q/ha.

ARKA SAURABH (SEL-4)

Arka Shurabh has been developed through selection from a breeding line (V-685) introduced from Canada at IIHR, Bangalore and identified for release through AICRP-VC in 1987. It is semi-determinate with broad light green foliage and good canopy. Fruits are thick, fleshy, firm, round, medium-large (70-75 g), deep red and with nipple tip, suitable for fresh market and processing and resistant to fruit cracking, good transport quality and shelf-life. It has been recommended for cultivation in J&K, H.P., Uttaranchal, Punjab, U.P., Bihar, Chhattisgarh, Orissa, A.P., Rajasthan, Gujarat, Haryana, Delhi, M.P., Maharashtra, Karnataka, Tamil Nadu and Kerala and yield potential of 350-370 q/ha in 140 days of crop duration.

SEL-1-6-1-4

Sel-1-6-1-4 is a root knot nematode resistant variety developed at PAU, Ludhiana and identified for release through AICRP-VC in 1987.

This variety is derived through selection from a cross between root knot resistant line (E-24-C-8) and susceptible variety (Punjab Chhuhara). It is determinate with dark green foliage, which protects fruits from sunburn. Fruits are pear-shaped and medium (40-50 g). It has been recommended for cultivation in J&K, H.P., Uttaranchal, West Bengal, Assam, Punjab, U.P., Bihar, Chattisgarh, Orissa and A.P and has yield potential of 560-580 q/ha

HYBRID TOMATO (ARTH-3)

ARTH-3 has been developed at Ankur Seeds Private Limited, Nagpur and identified for release through AICRP-VC in 1992. Plants are determinate with medium dark green leaves; fruits are squarish round, deep red with 4.5 to 5.0% TSS and moderately tolerant to drought; harvesting starts after 80-85 days and continue up to 160-180 day. It has been recommended for cultivation in Assam, West Bengal, Maharastra, Western parts of M.P., Karnatka, Tamil Nadu and Kerala and has yield potential of 880-900 q/ha.

HYBRID TOMATO (ARTH-4)

ARTH-4 has been developed at Ankur Seeds Private Limited. Nagpur and identified for release through AICRP-VC in 1992. It is semi-spreading and indeterminate with dark green leaves. Fruits are deep red, flat round and red flesh with 4-4.5% TSS and moderately tolerant to drought. Harvesting starts after 80-85 days and continues up to 170-190 days. It has been recommended for cultivation in Punjab, Bihar, U.P., Karnatka, Tamil Nadu and Kerala. It has yield potential of 1000-1020 q/ha.

GULMOHAR (MTH-6)

MTH-6 has been developed at MAHYCO Seeds, Jalna and identified for release through AICRP-VC in 1992. Plants are indeterminate bearing red, square fruits each weighing 90 g with 4.17% TSS. First picking starts at 60-65 days of transplanting. It has been recommended for cultivation in Maharashtra, Western Parts of M.P., Karnatka, Tamil Nadu and Kerala and has yield potential of 930-950 q/ha.

ARKA ALOK (BWR-5)

This determinate variety is resistant to bacterial wilt, developed at IIHR, Bangalore and identified for release through AICRP-VC in 1992. It is derived through selection from an introduced line (CL-114-5-1-0) from AVRDC, Taiwan. Fruits of early clusters are square-round, large (120 g) and fruits of later clusters are oblong, medium (80 g), firm with light green shoulder. This variety is suitable for kharif and rabi season, has been recommended for cultivation in West Bengal and Assam and has yield potential of 300-320 q/ha in 130 days of crop duration.

ARKA VARDAN (FMH-2)

Arka Vardan has been developed at IIHR, Bangalore and identified for release through AICRP-VC in 1992. Plants are indeterminate bearing flat round, firm, deep red fruits with green shoulder, each weighing 140 g. Plants are resistant to root knot nematode. It has been recommended for cultivation in J&K, H.P., U.P. Plains, Karnatka, Tami Nadu and Kerala and has yield potential of 550-600 q/ha.

SEL-32

Sel-32 is a determinate variety, developed at HAU, Hisar and identified for release through AICRP-VC in 1996. Fruits are flatish-round, smooth, large, attractive red and less-seeded, suitable for spring and summer season cultivation. It has been recommended for cultivation in West Bengal, Assam, Rajasthan, Gujarat, Haryana, Delhi, M.P. and Maharashtra and has yield potential of 320-340 q/ha.

DT-10

DT-10 has been developed at IARI, New Delhi and identified for release through AICRP-VC in 1996. Plants are indeterminate and erect with thick stem and green broad foliage. Fruits are medium, round, deep red with thick pericarp, suitable for processing. It has been recommended for cultivation in Punjab, U.P., Bihar, Rajasthan, Gujarat, Haryana and Delhi and has yield potential of 360-380 q/ha.

UTKAL URBASI (BT-12)

This variety has been derived from a cross (Punjab Chhuhara × LE-79) at OUA&T, Bhubaneshwar and identified for release through AICRP-VC in 1996. Plants are indeterminate, 70-85 cm tall, dense foliage, bearing 5-8 fruits/cluster. Fruits are red, pear-shaped, thick skin with green shoulder, self-life 10-12 days, 2-3 locules/fruit and less-seeded. First picking starts at 100-115 days of seed sowing. Plants are resistant to bacterial wilt and tolerant to early blight diseases. It has been recommended for cultivation in J&K, Himanchal Pradesh, Uttaranchal, Punjab, U.P. and Bihar and has yield potential of 380-400 q/ha.

H-24

H-24 is a determinate variety resistant to TLCV, developed through back cross-pedigree method from an inter-specific cross (Sel-7 × *L. hirsutum* f. *glabratum*) at IIVR and identified for release through AICRP-VC in 1997. Fruits are flatish-round, fleshy, medium and red. It has been recommended for cultivation in Chhattishagarh, Orissa and A.P. and has yield potential of 380-400 q/ha.

PUSA HYBRID-8 (DTH-8)

This variety has been developed at IARI, New Delhi and identified for release through AICRP-VC in 2001. Plants are determinate, with green compact foliage and heavy fruit bearer. Fruits are round, medium (75-80 g) with uniform ripening. It has been recommended for cultivation in Punjab, U.P., Bihar and Jharkhand and has yield potential of 430-450 q/ha.

UTKAL PRAGYAN (BT-116-3-2)

This variety has been developed at OUA&T, Bhubaneshwar and identified for release through AICRP-VC in 2001. Plants are determinate in growth habit bearing round, deep red, medium sized pulpy fruits with a blunt beak at distal end, and 5-6 fruits/cluster. Plants are tolerant to bacterial wilt disease. Fruits mature in 90 to 95 days after sowing. It has been recommended for cultivation in Chhattishgarh, Orissa, A.P., Rajasthan, Gujarat, Haryana and Delhi and has yield potential of 410-425 q/ha.

NARENDRA TOMATO-3 (NDT-3)

It is a determinate variety developed through hybridization (Resistant Bangalore × Kalyanpur Kuber) followed by pedigree selection at NDUA&T, Faizabad and identified for release through AICRP-VC in 2001. Fruits are small (40-50 g), flatish round, uniform red at maturity, suitable for processing, first picking starts at 60-65 days after transplanting. Plants are tolerant to root knot nematode and early blight. It has been recommended for cultivation in Punjab, U.P., Bihar, Jharkhand, Chhattishgarh, Orissa, A.P., Rajasthan, Gujarat, Haryana and Delhi and has yield potential of 350-380 q/ha.

AZAD T-6 (KS-118)

This variety has been developed at CSAUA&T, Kanpur and identified for release through AICRP-VC in 2001. Plants are determinate with early and prolific, smooth, compact fruiting, fruits round, red and medium sized (50 to 60 g). It has been recommended for cultivation in Punjab, U.P., Bihar and Jharkhand and has yield potential of 375-400 q/ha.

NARENDRA TOMATO-4 (NDT-9)

This indeterminate variety has been developed through hybridization (Pusa Ruby × Kalyanpur Kuber) followed by pedigree selection at NDUA&T, Faizabad and identified for release through AICRP-VC in 2001. Fruits are large (80-90 g), uniform red at maturity; flat round with good keeping and processing attributes. It has been recommended for cultivation in U.P., Bihar, Jharkhand and Uttaranchal and has yield potential of 350-375 q/ha.

UTKAL RAJA (BT-20-2-1)

This variety has been developed at OUA&T, Bhubaneshwar and identified for release through AICRP-VC in 2001. Plants have indeterminate growth habit with round, medium-large and deep red fruits. Fruits are in clusters of 4-5. Fruits get matured in 95 to 100 days after sowing. It is tolerant to bacterial wilt disease. It has been recommended for cultivation in Punjab, U.P., Bihar, Jharkhand, Madhya Pradesh and Maharashtra and has yield potential of 450-470 q/ha.

MEENAKASHI (BSS-20)

BSS-20 has been developed at Beejo Sheetal Seeds, Jalna and identified for release through AICRP-VC in 2001. Plants are indeterminate bearing dark red, round fruits each weighing 80-85 g and resistant to *Fusarium* wilt. It has been recommended for cultivation throughout the country and has yield potential of 1010-1020 q/ha.

KTH-2

This hybrid has been developed at CSAUA&T, Kanpur and identified for release through AICRP-VC in 2002. Plants are indeterminate bearing round and attractive red fruits. It has been recommended for cultivation in Punjab, U.P., Bihar, Jharkhand, Chhattisgarh, Orissa and A.P. and has yield potential of 550-600 q/ha.

NDTS-2001-3

This variety has been developed through hybridization (H-24 × NDT-120) followed by pedigree selection at NDUA&T, Faizabad and identified for release through AICRP-VC in 2004. Plants are determinate with heavy stem and dark green leaves. Fruits are oval round, solid, medium to large (60-80 g). First picking starts at 75-80 days after transplanting. Plants are highly tolerant to leaf curl virus and early blight. It has been recommended for cultivation in Punjab, U.P., Bihar and Jharkhand and has yield potential of 375-400 q/ha.

KTH-1

This hybrid has been developed at CSAUA&T, Kanpur and identified for release through AICRP-VC in 2004. Plants are determinate bearing 50 to 60 round, attractive red fruits per plant. It has been recommended for cultivation in Punjab, U.P., Bihar and Jharkhand and has yield potential of 550-600 q/ha.

LE-415

This variety has been developed at KAU, Vellanikkara and identified for release through AICRP-VC in 2004. Plants are semi-determinate bearing round, light red, medium size, slightly fasciated fruits free from green shoulder. Plants are resistant to bacterial wilt (*Ralstonia* Race1, biovar 3 & 5). It has been recommended for cultivation in J&K, H.P., Uttaranchal, Chhatishgarh, Orissa, A.P., Karnataka, Tamil Nadu and Kerala and has yield potential of 400-425 q/ha.

BT-136

This variety has been developed at OUAT, Bhubaneshwar and identified for release through AICRP-VC in 2005. Plants are indeterminate maturing in 95-98 days and bear round to oval, medium sized fruits with 4-5 fruits/cluster. Plants are tolerant to bacterial wilt. It has been recommended for cultivation in West Bengal, Assam, Punjab, U.P., Bihar and Jharkhandand has yield potential of 385-400 q/ha.

BRINJAL

Solanum melongena
Family : Solanaceae
Chromosome no.: 2n=24

Brinjal (syn: eggplant, aubergine) is grown on commercial scale in India. The other countries cultivating brinjal are China, Turkey, Japan, Egypt, Italy, Indonesia, Iraq, Syria, Spain, Philippines, Bulgaria and USA. In India, Orissa ranks first in terms of area under brinjal cultivation, while West Bengal ranks first in terms of total production. Bihar, Karnataka, Maharashtra, Andhra Pradesh and Uttar Pradesh are major brinjal growing states. It is highly productive and usually finds its place as poor mans vegetable, however, now it has occupied a place of prominence in day-to-day diet of all classes of society. India is considered to be the centre of origin of the eggplant due to occurrence of different names in Sanskrit such as 'Vartta Baigan' and bong, etc. The unripe, tender and soft fruits are mainly used as cooked vegetables for preparation of curry, while the long ones are used as fried or stuffed. It has got much potential as raw material in pickle making and dehydration industries. It is reported to stimulate the intrapeptic metabolism of blood cholesterol.

Brinjal plants are semi-erect to erect, herbaceous and branching in nature. Leaves are opposite or sub-opposite, large, ovate, shallowly sinuate, lobbed, and glabrous. Generally, plant has pentamerous and bears hermaphrodite flowers. Calyx is wooly or non-wooly, spiny or non-spiny, persistent and accrescent. Corolla is lobed with different colours. Stamens are free, erect, yellow with short filaments, flat at the base with long narrow anthers. Ovary is bilocular with multi-ovules. Fruit is berry with different colours. It is primarily a self-pollinated crop, but cross-pollination also occurs even if plants are heterostyly. Brinjal is a day-neutral plant, anthesis of the flowers usually takes place at 6-8 hours in summer and it is delayed upto 11 hours in winter. The optimum temperature for normal growth and development of plant and fruit is 21-30°C, whereas optimum temperature for seed germination is 21-25°C with 50-60% relative humidity. Brinjal can be grown in different types of soil, but well-drained fertile clay loam or silt loam with soil pH of 5.5-6.5 is the best.

In northern parts of India, nursery is normally raised in June-July (Autumn-Winter crop) or November (spring summer crop). Sowing time in other part of country is from June to September and again in December to January. In hilly areas, March-April is the best time for nursery raising. Seedlings become ready for transplantation in about 4-5 weeks depending upon the temperature. About 250-350 g seeds are required for planting 1 ha area.

The major insect problem in eggplant is shoot and fruit borer, which causes ~40% yield loss. Damping-off is a major disease of nursery, whereas little leaf, wilt and phomopsis blight considerably affect the crop in the field.

Varieties

In India, ICAR Institutes and SAUs are working for the genetic improvement of brinjal. Selection, hybridization and heterosis breeding primerily have been used for improvement of brinjal and several varieties/hybrids have been released and recommended for cultivation under different agro-climatic zones as described below (for photographs see plate 2).

PUSA PURPLE LONG (PPL)

This variety has been developed through single plant selection from a cultivar (Batia) at IARI, New Delhi and notified by CVRC Notification nos. 716, 540(E), 361(E), 440, and 786, dated, 20.2.1970, 20.07.1985, 30.6.1973, 21.08.1975, and 02.02.1976, respectively. Fruits are long (20-25 cm), purple, glossy and tender. It is suitable for summer and autumn season cultivation. It has been recommended for cultivation in Punjab, U.P., Bihar, Rajasthan, Gujarat, Haryana, Delhi, M.P., Maharashtra, Karnataka, Tamil Nadu and Kerala and has yield potential of 280-310 q/ha.

PUSA PURPLE ROUND (PPR)

This variety has been developed at IARI, New Delhi and notified by CVRC Notification nos. 716, 440, 786, 13, and 540(E), dated 20.2.1970, 21.08.1975, 02.02.1976, 19.12.1978, and 20.07.1985, respectively. Plants are very tall with a thick greenish purple stem and highly serrated deep green leaves. On an average, each fruit weigh ~140 g with a bearing of 6 fruits per plant. It is resistant to shoot and fruit borer and little leaf disease and has yield potential of 275-300 q/ha.

BLACK BEAUTY

This variety is an introduction from USA and notified by CVRC Notification nos. 440, and 13, dated 20.2.1970, and 02.02.1976, respectively. Plants are 60-75 cm tall, erect and sturdy with dark green leaves. Its flowering starts at 80-90 days after transplanting. Fruits are roundish oblong, 10-12 cm long with dark purplish black skin. This variety has yield potential of 250-270 q/ha.

PUSA PURPLE CLUSTER (PPC)

This early variety has been derived through single plant selection from an indigenous collection at IARI, Regional Station, Katrain and notified by CVRC Notification nos. 440(E) and 1004 dated 21.08.1975 and 23.03.1978, respectively. Deep purple fruits borne in cluster of 4-9 and 10-12 cm in length. It is moderately resistant to bacterial wilt. It has been recommended for cultivation in Punjab, U.P., Bihar, Orissa, A.P., Rajasthan, Gujarat, Haryana and Delhi and has yield potential of 300-325 q/ha.

H-4

This variety has been developed through hybridization (Hyderpur × PPL) followed by pedigree selection at HAU, Hisar and notified by CVRC Notification no. 786, dated 02.02.1976. It fruits are medium to long, thin and dark purple with light green flesh. It has yield potential of 225-250 q/ha.

PUNJAB HARYANA BRINJAL-4

This variety has been named as Punjab Haryana Brinjal-4 for being a derivative of the cross Hyderpur × Pusa Purple Long under the selection no. S-4 done at PAU, Ludhiana and selection no. H-4 done at HAU, Hisar and notified by CVRC Notification no. 1004 dated 23.03.1978. Both of the genotypes are identical in major agronomic and economic characters except that S-4 has long and thin fruits, whereas H-4 has comparatively shorter fruits. Plants are bushy with pigmented stem. Its fruits are long to medium, thin, dark purple with light green flesh. It has been recommended for general cultivation in Punjab, Haryana, Uttar Pradesh, Delhi, Rajasthan and Gujarat and has yield potential of 225-250 q/ha.

JAMUNI GOLA (S-16)

This variety has been developed at PAU, Ludhiana and notified by CVRC Notification no. 165 (E) dated 06.03.1987. Plants are medium in height, spreading, thornless and early maturing. Fruits are oval-round, plump and shining purple. First picking starts 65 days after transplanting. Suitable for cultivation in autumn season and recommended for cultivation for Chhattisgarh, Orissa and A.P., Rajasthan, Gujarat, Haryana and Delhi and has yield potential of 250-270 q/ha.

PUNJAB BAHAR

This variety has been developed at PAU, Ludhiana and notified by CVRC Notification no. 13 dated 19.12.1978. It is a selection from local material from Punjab. Plants are ~95 cm tall and non-spiny with deep purple foliage. First picking starts 70 days after transplanting. Its fruits are round, dark purple, shining, weighing about 200-300 g and less-seeded and suitable for Bhartha making. It has yield potential of 275-300 q/ha.

PUNJAB CHAMKILA

This variety has been developed at PAU, Ludhiana and notified by CVRC Notification no. 13 dated 19.12.1978. Fruits are long, thin and dark purple. It is suitable for summer and autumn season and has yield potential of 430-450 q/ha.

BHAGYAMATI

This variety has been developed through selection from a local germplasm at ANGRAU, Hyderabad and notified by CVRC Notification no. 19(E) dated 14.01.1982. Plants are 50-70 cm tall with tender stem and purple shoots. Emerging and mature leaves are purple. Branching begins at an early stage and plant appears compact Fruits are deep shining purple, oblong, 6-8 cm in length and borne in cluster of 2-6. Calyx five and has a green to purplish hue in different seasons. It has yield potential of 150-180 q/ha in 150-170 days of crop duration.

MDU-1

This variety is a selection from Kallampatti (a genotype of Madurai) has been developed at TNAU, Coimbatore and notified by CVRC Notification no. 19(E), dated 14.01.1982. The variety is adaptable to tropical plain of Tamil Nadu. Its plants are compact and medium spreading. Leaves are broad, green without pigments. Fruits are bright purple, round and each of 280 g in weight. As the fruit matures, the purple colour fades to pale pink. It has yield potential of 340-350 q/ha in 135-145 days of crop duration.

KALYANPUR T-3

This variety has been developed at CSAUA&T, Kanpur and notified by CVRC Notification no. 2(E), dated 03.01.1983. Fruits are round, light, purple and whitish green at stigmatic end. It has been recommended for cultivation in Punjab, U.P., Bihar, Rajasthan, Gujarat, Haryana, Delhi, Karnataka, Tamil Nadu, and Kerala and has yield potential of 300-325 q/ha.

JUNAGADH LONG

This variety is a derivative of the cross (Doli-5 × Panch Mahal) developed at GAU, Junagadh and notified by CVRC Notification no. 596(E), dated 13.08.1984. Plant height is 120 cm with semi-spreading growth habit. Leaves are dark green with wavy margins. Plants are non-spiny. Fruits are medium in size, tender, pinkish purple, long (~17 cm) with a girth of ~8 cm at the centre and tapering at the distal end. 20-25 pickings can be taken at short intervals. It is free from collar rot disease under field conditions and has yield potential of 300-350 q/ha.

GUJARAT OBLONG BRINJAL- 1

This variety has been developed through selection from a local cultivar at AAU, Anand and notified by CVRC Notification no. 597 (E) dated 25.04.2006. Fruits are medium-sized, oblong and attractive due to shining dark black. It is moderately tolerant to little leaf disease. It is recommended for central Gujarat and Saurashtra region of the Gujarat and has yield potential of 470-490 q/ha.

AZAD KRANTI (KS-701)

This variety is derived from a cross (Banarasi Giant Long × Pusa Purple Long) developed at CSAUA&T, Kanpur and notified by CVRC Notification no. 295(E) dated 09.04.1985. Plants are medium-tall, profusely branched with semi-erect stem having dark green wavy margin and bear 12-15 fruits. Fruits are medium, thick-long, smooth, taper at distal end, dark purple with green calyx and white flesh, weight ranges from 100 to110 g; first picking starts 60-65 days after transplanting; suitable for both late and early sown conditions. It is tolerant to fruit rot, little leaf, drought and frost. It has been recommended for cultivation in Punjab, U.P., and Bihar and has yield potential of 310-330 q/ha in a crop duration of 140-150 days.

PANT SAMRAT

This variety has been derived through selection from a local collection at GBPUA&T, Pantnagar and notified by CVRC Notification no. 295(E) dated 09.04.1985. It is resistant to phomopsis blight and bacterial wilt and suitable for rainy season cultivation. It has been recommended for cultivation in Punjab, U.P., Bihar, Chhattisgarh, Orissa, A.P., M.P. and Maharashtra and has yield potential of 390-415 q/ha.

PANT RITURAJ (PBR-91-2)

This variety has been developed through pedigree method from a cross (Type-3 × Pusa Purple Cluster) at GBPUA&T, Pantnagar and notified by CVRC Notification no. 540(E) dated 24.07.1985. Plants are semi-erect with dark green foliage and prolific bearing. Its fruits are round, purple, slightly tapering towards the bottom, first picking starts 60 days after transplanting. It is suitable for planting in winter and summer seasons. It has been recommended for cultivation in Punjab, U.P., Bihar, Rajasthan, Gujarat, Haryana, Delhi, M.P., Karnataka, Tamil Nadu and Kerala and has yield potential of 310-325 q/ha.

CO-1

This variety is a pure line selection developed at TNAU, Coimbatore and notified by CVRC Notification no. 540(E) dated 24.07.1985. Plants are erect, compact and bushy with green stem and leaves and greenish purple petiole. The fruits are oblong and medium-sized with pale green shade under white background. Each fruit on an average weigh 50 to 60 g with good keeping quality. The seeds are soft even at full maturity. It is moderately resistant to root knot and reniform nematodes. It is preferred in the southern districts of Tamil Nadu and has yield potential of 240-260 q/ha in a crop duration of 160 days.

ANNAMALAI

This variety is a selection from the germplasm AC-49A, developed at TNAU, Coimbatore and notified by CVRC Notification no. 832(E) dated 18.11.1985. Plant height is 65.3 cm with a spread of 74.3 cm. Plant semi-erect characterized by disposition of young leaves. In winter, the fruits are deep purple, whereas lighter in summer and suitable for ratooning. Its fruits are somewhat club-shaped and slightly curved in the middle. It is resistant to aphids and has yield potential of 310-325 q/ha.

ARKA SHIRISH

This variety is an improvement over the *Irangeri Brinjal* of Karnataka, developed at IIHR, Bangalore and notified by CVRC Notification no. 867(E) dated 26.11.1986. Its fruits are very tender, extra-long, green, thick with attractive light green skin, good edible flesh and less seeds. The flesh texture is good with better cooking quality. Each plant bears 25 to 30 fruits. It has yield potential of 390-410 q/ha.

ARKA SHEEL

This variety has been developed at IIHR, Bangalore and notified by CVRC Notification no. 867(E) dated 26.11.1986. It is derived through single plant selection from a local collection (Bangalore-192) from Kodagn of Karnataka. Plants are tall with green leaves, purple tinge at the base in young leaves and solitary bearing habit. Fruits are medium-long, deep shining purple skin with green calyx. It has been recommended for cultivation in the states of Karnataka, Tamil Nadu and Kerala and has yield potential of 380-400 q/ha in 150-160 days of crop duration.

ARKA NAVNEET

This variety has been developed at IIHR, Bangalore and notified by CVRC Notification no. 867(E) dated 26.11.1986. Plants are indeterminate. Its fruits are deep purple, gossy black, each weighing 350-400 g with green calyx, flesh white with few seeds and good cooking and keeping quality. It has yield potential of 700-710 q/ha in 120 days of crop duration.

PUNJAB JAMUNI GOLA

This variety has been developed at PAU, Ludhiana and notified by CVRC Notification no. 1004 dated 23.03.1978. It is early maturing and suitable for autumn season sowing. Fruits are oval-round and purple. This variety has yield potential of 300-320 q/ha.

PUNJAB BARSATI

This variety has been developed through hybridization (PPC x H-4) followed by pedigree selection at PAU, Ludhiana and notified by CVRC Notification no. 280(E) dated 13.04.1989. Plants are dwarf, erect and thornless with green foliage. Fruits are medium-long, shining purple and tolerant to fruit borer. It is suitable for rainy season cultivation. First picking starts at 60-65 days after transplanting. It has been recommended for cultivation in Punjab, U.P. and Bihar and has yield potential of 360-380 q/ha.

PUNJAB SADABAHAR

This variety has been developed through hybridization (Japanese Long × R-34) followed by pedigree selection at PAU, Ludhiana and notified by CVRC Notification no. 13 dated 19.12.1978. Its plants are dwarf, erect and thornless. Fruits are long, thin and deep purple. It is good for summer and autumn season cultivation and also as ratoon crop. It is tolerant to shoot and fruit borer and has yield potential of 300-320 q/ha.

ARUNA

This variety has been developed through selection from a local cultivar at PDKV, Akola and notified by CVRC Notification no. 915(E) dated 06.11.1989. Plants are prolific bearing with small, roundish-oval, bright plum fruits with red skin. First picking starts 95 days after transplanting. It has been recommended for cultivation in M.P. and Maharashtra and has yield potential of 415-430 q/ha and 220-240 q/ha in rainy and summer season, respectively.

PUSA ANUPAM

This hybrid has been developed through heterosis breeding (Pusa Kranti x Pusa Purple Cluster) at IARI, Regional Station, Katrain and notified by CVRC Notification no. 527(E) dated 16.08.1991. The plants are medium in size and bushy. This variety takes 135-145 days from sowing to first picking. Fruits are borne in cluster of 3-5. Fruits measure 14-18 cm length, 14-15 cm girth and 3.4-4.5 cm in diameter. Fruits are cylindrical, medium-long, tender, firm and purple in colour. Fruit are also suitable for pickle making. It is available from August to November in the hilly regions. It has yield potential of 300-400 q/ha with 30-35 fruits/plant.

UTKAL TARINI (BB-7)

This variety has been developed at OUA&T, Bhubaneshwar and notified by CVRC Notification no. 860(E) dated 25.11.1992. Plants are medium in height. Fruits are oblong, medium-sized, deep purple, solitary bearing and sometimes two fruits per cluster. Fruit picking starts in 90 to 95 days after sowing. It is resistant to bacterial wilt and has yield potential of 345-365 q/ha.

PUSA HYBRID-6

This hybrid has been developed at IARI, New Delhi and notified by CVRC Notification no. 617(E) dated 17.08.1983. Plants are non-spiny, semi-erect, branched, vigorous, partial pigmentation on younger leaves and growing tips. Fruits are round, glossy, attractive, violet purple, partially pigmented peduncle with each fruit weighing 200-250 g. Being an early bearing hybrid it takes 85-90 days from sowing to first picking and further pickings continue up to December. It has yield potential of 450-470 q/ha.

HISAR SHYAMAL (H-8)

This variety has been developed at HAU, Hisar and notified by CVRC Notification no. 617(E) dated 17.08.1983. It is derived through backcross-pedigree method from the cross Aushey × BR-112. Plants are spreading, with purple flower, green leaves at seedling stage, which turns purple at maturity. Fruits are round, dark bright purple with creamish white flesh. First picking starts 50-60 days after transplanting. It has been recommended for cultivation in West Bengal, Assam, Punjab, U.P., Bihar, Chhatisgarh, Orissa, A.P, Rajasthan, Gujarat, Haryana, Delhi, M.P., Maharashtra, Karnataka, Tamil Nadu and Kerala and has yield potential of 330-350 q/ha in 150 days of crop duration.

HISAR PRAGATI (H-7)

This variety has been developed through hybridization (R-34 x Sel-26) followed by pedigree selection at HAU, Hisar and notified by CVRC Notification no. 636(E) dated 02.09.1994. Plants are compact, dwarf, bushy, indeterminate and erect with green foliage at early stage, which turns purple at later stage. Flowers are whitish purple with green sepals and peduncle. Fruits are long, dark bright purple with creamish white flesh. First picking starts at 45-50 days after transplanting. Plants are tolerant to little leaf disease and high temperature. It has been recommended for cultivation in Punjab, U.P., Bihar, Rajasthan, Gujarat, Haryana and Delhi and has yield potential of 340-360 q/ha in 150 days of crop duration.

PUSA HYBRID-5

This hybrid has been developed at IARI, New Delhi and notified by CVRC Notification no. 636(E) dated 02.09.1994. Plants are vigourous, non-spiny with semi-erect branches. Fruits are long, glossy, attractive, dark purple with partially pigmented peduncle, each fruit weighing 100 g. It takes 80-85 days from sowing to first picking and further pickings continue up to December in the Central and Northern Plains, whereas in Kerala, Tamil Nadu, Karnataka and Andhra Pradesh from November to May. It has been recommended for cultivation in northern plains, central India, non-coastal areas of Kerala, Tamil Nadu, Karnataka and Andhra Pradesh and has yield potential of 520-540 q/ha.

PANT BRINJAL HYBRID-1

This hybrid has been developed through heterosis breeding (PB-129 × PB-225) at GBPUA&T, Pantnagar and notified by CVRC Notification no. 636 (E) dated 02.09.1994. Plants are medium with purplish green stem. The leaves are dark green with purple tinge. Fruits are long, bright deep purple with soft flesh. Plants take 70-75 days for first picking. It is resistant to bacterial wilt and Phomopsis blight under field conditions and has yield potential of 600-620 q/ha.

ANAND BRINJAL HYBRID-1 (ABH-1/GBH-1)

This hybrid has been developed through heterosis breeding (M-2 × M-35) at GAU, Anand and notified by CVRC Notification no. 115(E) 10.2.1996. Fuits are dark purple with single green styler spot at apex. It is tolerant to shoot and fruit borer as well as little leaf disease. It has been recommended for cultivation in Punjab, U.P., Bihar, Jharkhand, Rajasthan, Gujarat, Haryana and Delhi, Madhya Pradesh and Maharashtra and has yield potential of 560-580 q/ha

BH-1

This hybrid has been developed through heterosis breeding (Punjab Barsati × Jamuni Gola) at OUA&T, Bhubaneshwar and notified by CVRC Notification no. 115(E) dated 10.02.1996. Plants are medium in height, compact and thornless. Fruits are oblong, plumpy and shining purple. It is early in fruiting and first picking starts after 50-55 days of transplanting. It is tolerant to fruit borer, can be planted in February and August and has yield potential of 510-530 q/ha.

BH-2

This hybrid has been developed through heterosis breeding (Punjab Barsati × Punjab Neelam) at OUA&T, Bhubaneshwar and notified by CVRC Notification no. 115(E) dated 10.02.1996. Its leaves are green and purplish. Plants are medium, erect, spreading and thornless. Its fruits are oblong and deep purple. Average fruit weight is 300 g. It is highly suitable for Bharta making, tolerant to fruit borer and has yield potential of 515-530 q/ha.

NARENDRA BRINJAL-1 (NDB-25)

This variety has been developed at NDUA&T, Faizabad and notified by CVRC Notification no. 115(E) dated 10.02.1996. Plants are long, erect with greenish purple leaves. Fruits are dark purple, soft, less-seeded and oblong. Fruits are ready for harvest in 70-75 days after trans planting. This variety has yield potential of 300-350 q/ha.

NARENDRA HYBRID BRINJAL-3 (NDBH-3)

This hybrid has been developed at NDUA&T, Faizabad and notified by CVRC Notification no. 597(E) dated 25.04.2006. Plants are spreading type with light purple leaves. Fruits are round, large-sized, soft and shining purple and are ready to harvest in 75-80 days after planting. It has yield potential of 500-550 q/ha.

PUSA UTTAM

This variety has been developed at IARI, New Delhi and notified by CVRC Notification no. 98(E) dated 08.02.1997. Plants are semi upright, vigorous, branched and free from spines. Mature plants appear green with occasional light pigmentation also present on growing shoots. Flowers appear in clusters having bluish-violet corolla. Fruits are pendent, oval, medium, large-sized, and glossy with dark purple skin and green peduncle. The bearing habit is solitary. Single fruit weight 250-300 g. It has yield potential of 425-450 q/ha.

PUSA UPKAR (DBR-8)

This variety has been developed at IARI, New Delhi and notified by CVRC Notification no. 98(E) dated 08.02.1997. Plants are semi-upright, medium-tall (70 cm) with 5-6 branches and greenish purple leaves bearing ~15 fruits. Fruits are round (9 cm diameter), dark purple and each fruit weighing ~300 g. It has been recommended for cultivation in Punjab, U.P. and Bihar and has yield potential of 650-670 q/ha.

PUSA BINDU (DBRS-44)

This variety has been developed at IARI, New Delhi and notified by

CVRC Notification no. 98(E) dated 08.02.1997. Plants are ~60 cm tall with green leaves. Fruits are borne in cluster of 4-5 with each fruit weighing 125 g. This variety has yield potential of 250-300q/ha.

UTKAL MADHURI (BB-44)

This bacterial wilt resistant variety has been developed through hybridization (PBR-125-5 × Pipili-4) followed by pedigree selection at OUA&T, Bhubaneshwar and notified by CVRC Notification no. 834(E) dated 21.01.1998. Plants are of medium stature with green stem and leaves, bear solitary fruits. Fruits are green (basal portion white), long, medium with white flesh having 7-8 days of shelf-life. It is resistant to *Fusarium* wilt and drought. First picking starts 100-105 days after transplanting. It has been recommended for cultivation in M.P., Orissa, A.P., and Maharashtra and has yield potential of 320-340 q/ha in 160 days of crop duration.

UTKAL KESHARI (BB-26)

This variety has been developed through hybridization (BB-11 × KJ-3-1) followed by pedigree selection at OUA&T, Bhubaneshwar and notified by CVRC Notification no. 834(E) dated 21.01.1998. Plants are tall, erect with purple pigmentation on branches and leaves. Fruits are long, medium-large, uniform thick, basal portion slightly broad and deep purple with 5-6 days of shelf-life. It is tolerant to bacterial and *Fusarium* wilt. It has been recommended for cultivation in Rajasthan, Gujarat, Haryana, Delhi, M.P. and Maharashtra and has yield potential of 325-340 q/ha in 130 to 140 days of crop duration.

JC-1

This variety has been developed at JNKV, Jabalpur and notified by CVRC Notification no. 834 (E) dated 21.01.1998. Plants are tall with non-spiny pedicel and elongated, medium-sized purple fruits and pointed apex. It is resistant to phomopsis blight and little leaf virus, wilt and fruit borer and has yield potential of 400-450 q/ha.

JC-2

This variety has been developed at JNKV, Jabalpur and notified by CVRC Notification no. 834 (E) dated 21.01.1998. Plant are tall with spiny pedicel and bear large, oblong purple fruits with blunt apex, no incidence of little leaf, phomopsis blight, wilt and fruit borer infestation. This variety has yield potential of 480-500 q/ha.

SWETHA (SM 6-6)

This bacterial wilt resistant variety has been developed at KAU, Vellanikkara and notified by CVRC Notification no. 834 (E) dated 21.01.1998. Plants are bushy with purple tinge on leaf stalk. Fruits are medium-long, solitary (occasionally in cluster). It can be grown in seasons of May-August, September-December and January-March. It has been recommended for cultivation in J&K, H.P., Uttaranchal, Rajasthan, Gujarat, Haryana, Delhi, Karnataka, Tamil Nadu and Kerala and has yield potential of 300-325 q/ha in 135-150 days of crop duration.

DBSR-91

This variety has been developed at IARI, New Delhi and notified by CVRC Notification no. 1052(E) dated 26.10.1999. Plants are semi-upright (semi-erect stem), medium-tall (65 cm) with 6-8 branches. Fruits are small, round and purple. It has been recommended for cultivation in M.P. and Maharashtra and has yield potential of 560-580 q/ha.

AZAD B-1

This variety has been developed at CSAUA&T, Kalyanpur, Kanpur and notified by CVRC Notification no. 295(E) dated 09.04.1985. It is a derivative of a cross (Banarasi Giant Round × 7103) suitable for July-August transplanting in north and north-western regions. Plants are dwarf (60-70 cm), stout, erect and compact. Leaf lamina is green, wide and thick. Fruits are smooth, round and shining purple. The average fruit weight at edible stage is 160-180 g and mean fruit number per plant is 10-12. It is moderately resistant to little leaf, phomopsis fruit rot and blight and less-susceptible to shoot and fruit borer. It has yield potential of 1.5-2 q/ha and 300-350 q/ha seed and marketable fruit, respectively.

PKM-1

This variety has been developed at TNAU, Coimbatore and notified by CVRC Notification no. 1052(E) dated 26.10.1999. It is a pure line selection from Kulathur Local near Tirunelveli. It is suitable for cultivation in both irrigated and rainfed conditions. Egg-shaped and milky white fruits are borne in cluster (2-4). The crop duration is 130-135 days. It is preferred in the markets of southern districts of Tamil Nadu and has yield potential of 360-380 q/ha.

SWARNA MANI (CHBR-1)

This variety has been developed at HARP, Ranchi and notified by CVRC Notification no. 642 (E) dated 31.05.2004. Plants are prostrate (spreading stem), medium-tall (89.0 cm) with 6-9 branches and violet leaves, bearing 15 round (9.6 cm), dark violet fruits each weighing 320 g. It has been recommended for cultivation in Punjab, U.P. and Bihar and has yield potential of 750-770 q/ha.

SWARNA PRATIBHA (CH-309)

This bacterial wilt resistant variety has been developed through pure line selection at HARP, Ranchi and notified by CVRC Notification no. 642 (E) dated 31.05.2004. First harvest can be taken 75-80 days after planting. Fruit are medium-sized (250 g), and shiny purple. It has been recommended for cultivation in Bihar and Jharkhand and has yield potential of 470-500 q/ha.

UTKAL JYOTI (BB-13)

This variety has been developed through hybridization (KT-4 × BB-11) followed by pedigree selection at OUA&T, Bhubaneshwar and notified by CVRC Notification no. 642 (E) dated 31.05.2004. Plants are medium-tall, non-spiny and bear profuse fruits in clusters. Fruits are long, small-medium, purple with white flesh and better shelf-life. It is tolerant to bacterial wilt. It has been recommended for cultivation in Karnataka, Tamil Nadu and Kerala and has yield potential of 350-375 q/ha in 140-180 days of crop duration.

KASHI PRAKASH (IVBR-1)

This oblong fruited variety has been developed at IIVR, Varanasi and notified by CVRC Notification no. 597(E) dated 25.04.2006. Plants are semi-upright with green stem and leaves. Fruits are attractive with light green spotted, calyx spiny and each weighing ~190 g. It has been recommended for cultivation in Punjab, U.P., Bihar and Jharkhand and has yield potential of 650-675 q/ha.

KASHI TARU (IBVL-9)

This variety has been developed through mass selection at IIVR, Varanasi and notified by CVRC Notification no. 597(E) dated 25.04.2006. It is tall with erect growth habit having dark green leaves and stem. Plant height 120-130 cm bearing 20-25 fruits/plant. Fruit picking starts 75-80 days after transplanting. Fruits are long (31 cm), dark purple, glossy and tender and each fruit weighing 745-85 g. It is suitable for summer and autumn season cultivation and has yield potential of 700-720 q/ha.

PHULE HARIT (RHRB-16)

This variety has been developed through selection from a germplasm at MPKV, Rahuri and notified by CVRC Notification no. 2035(E) dated 28.11.2006. Plants are tolerant to fruit borer and little leaf under field conditions. Fruits are smooth and attractive green with white strips at apical end and bigger in size. It has yield potential of 350-370 q/ha in 180 days of crop duration.

SWARNA SHYAMALI

This variety has been developed through pure line selection at HARP, Ranchi and notified by CVRC Notification no. 2035(E) dated 28.11.2006. Fruit are medium-sized (250 g), round, attractive green with white stripes. Plants are resistant to bacterial wilt. First harvest can be taken at 45-60 days after planting. It has been recommended for cultivation in Bihar and Jharkhand and has yield potential of 500-525 q/ha.

SWARNA SHREE

This variety has been developed through pure line selection at HARP, Ranchi and notified by CVRC Notification no. 2035(E) dated 28.11.2006. Fruits are creamy white, round to oval, soft and preferred for *bharta* preparation. Plants are moderately resistant to bacterial wilt. It is suitable for July-August sowing and first harvest can be taken 55-60 days after planting. It has been recommended for cultivation in Bihar and Jharkhand and has yield potential of 550-600 q/ha.

SWARNA AJAY (HABH-3)

This hybrid has been developed through heterosis breeding at HARP, Ranchi and notified by CVRC Notification no. 2035(E) dated 28.11.2006. Fruit are attractive, light purple, oblong, medium length (10-12 cm) and each fruit weighing 100-120 g. First harvest can be taken 55-65 days after planting. Plants are resistant to bacterial wilt and phomopsis blight. This hybrid has been recommended for cultivation in Punjab, Uttranchal, Bihar, Jharkhand and U.P. and has yield potential of 700-720 q/ha.

KASHI SANDESH (VRBHR-1)

This hybrid has been developed at IIVR, Varanasi and notified by CVRC Notification no. 2035(E) dated 28.11.2006. Plants are semi-upright with green stem and purplish-green leaves and purple flowers. Fruits are round, light purple, medium-sized, average fruit weight is 225 g, First picking starts at 75 days after transplanting. This hybrid has been recommended for cultivation in Punjab, U.P., Bihar, Jharkhand, Rajasthan, Gujarat, Haryana and Delhi and has yield potential of 800-820 q/ha.

ARKA KESHAV (IIHR-21)

This variety has been developed through hybridization (Dingrass Multiple Purple × Arka Sheel) followed by pedigree selection at IIHR, Bangalore and notified by CVRC Notification no. 2035(E) dated 28.11.2006. Plants are tall, branched, light purple green stem with green leaves bearing long fruits in clusters. Fruits tender with slow seed maturity and free from bitter principles. Plants are resistant to bacterial wilt and has yield potential of 450-470 q/ha in 150 days of crop duration.

ARKA NIDHI (BWR-12)

This variety has been developed through hybridization (Dingrass Multiple Purple × Arka Sheel) followed by pedigree selection at IIHR,

Bangalore and notified by CVRC Notification no. 2035(E) dated 28.11.2006. Plants are tall, compact with purple green stem and veins and bears fruit in cluster. It bears medium-long, tender fruits with blue-black, glossy green purple calyx. Plants are resistant to bacterial wilt. This variety has been recommended for cultivation in Karnataka, Tamil Nadu and Kerala and has yield potential of 485-500 q/ha in 150 days of crop duration.

NARENDRA HYBRID BRINJAL-1 (NDBH-1)

This hybrid has been developed through heterosis breeding (PBR-91-1 × K202-14-90-110) at NDUA&T, Faizabad and notified by CVRC Notification no. 115(E) dated 10.02.1996. Plants are semi-erect, non-spiny, hardy having purple green leaves with round, purple, attractive, shining, soft, fleshy, less-seeded fruits with good flavour and each weighing 150-200 g. It can be harvested 70-75 days after transplanting. Plants are moderately resistant to little leaf disease. This variety has been recommended for cultivation in Punjab, Bihar, U.P., Haryana, Rajasthan, Gujarat, Maharashtra, and M.P. and has yield potential of 450-500 q/ha.

JUNAGADH OBLONG

This variety has been developed at GAU, Junagadh and notified by CVRC Notification no. 596(E) dated 13.08.1984. Fruits are oblong and purple coloured. It is moderately resistant to little leaf disease and fruit and shoot borer and has yield potential of 280-310 q/ha.

GBL-1

This variety has been developed at GAU, Anand and notified by CVRC Notification no. 448 (E) dated 21.04.2003. Plants are spreading type with light purple and medium long fruits. This variety has yield potential of 320-340 q/ha.

GUJRAT BRINJAL HYBRID-2 (GBH-2)

This hybrid has been developed through heterosis breeding (JBGL-81 × JBGL-51) at GAU, Anand and notified by CVRC Notification no. 642(E) dated 31.05.2004. Plants are erect. Fruits are round, blackish purple with green calyx. This hybrid has been recommended for cultivation throughout the country and has yield potential of 350-375 q/ha.

PUSA KRANTI

This variety has been derived from a three way cross (PPL x Hyderpur) × Wynad Giant was developed at IARI, New Delhi and identified for release through AICRP-VC in 1977. Fruits are oblong, 15-20 cm long, dark purple with shining green calyx and less-seeded. It has been recommended for cultivation in Punjab, U.P. and Bihar and has yield potential of 275-300 q/ha.

PUSA ANUPAM (KAT-4)

This variety has been developed through pedigree method at IARI Regional Station, Katrain and identified for release through AICRP-VC in 1987. Fruits are of medium size, cylindrical, tender, firm and purple borne in clusters of 3-5. It is resistant to wilt, has been recommended for the cultivation in Karnataka, Tamil Nadu and Kerala and has yield potential of 370-400 q/ha.

ARU-2C

This is a bacterial wilt resistant variety, developed through selection at DARL, Pithoragarh and identified for release through AICRP-VC in 1987. Fruits are cylindrical, long, bright, violet, borne in cluster of 4-6. It is recommended for cultivation in J&K, H.P., Uttaranchal, Punjab, U.P., Bihar, Rajasthan, Gujarat, Haryana, Delhi, M.P. and Maharashtra and has yield potential of 310-325 q/ha.

UTKAL TARINI (BB-7)

This bacterial wilt resistant variety has been developed through hybridization (Pusa Kranti × Gopa Local) followed by pedigree selection at OUA&T, Bhubaneshwar and identified for release through AICRP-VC in 1990. Plants are of medium stature with purple pigmentation on leaves and stem bearing solitary fruit (occasionally, two fruits/ cluster). Fruits are oblong, medium-sized, deep purple with creamish white flesh with 6-7 days shelf-life. First picking starts 90-95 days after transplanting. It is recommended for cultivation in West Bengal,

Assam, Chhattisgarh, Orissa and A.P. and gives an average yield of 340 q/ha in 130-140 days (after transplanting) of crop duration.

SURYA (SM-6-7)

Surya is a bacterial wilt resistant variety, developed at KAU, Vellanikkara, and identified for release through AICRP-VC in 1992. Plants are non-prickly and spreading bearing oval and purple fruits. This variety has been recommended for the cultivation in Karnataka, Tamil Nadu and Kerala and has yield potential of 200-225 q/ha in 150-180 days of crop duration.

KALPATARU (MHB-10)

MHB-10 has been developed through heterosis breeding (MBA-113 x MBR-129) at MAHYCO Seeds, Jalna and identified for release through AICRP-VC in 1993. Plants are erect, bushy, compact and spiny. Fruits are round to slight oval, shining, redish purple with white streaks (bicolour), green spiny calyx with fruit weight of 60-70 g. First picking starts 70-75 days after transplantation. This variety has been recommended for cultivation in Punjab, U.P., Bihar, Jharkhand, Chhatishgarh, Orissa, A.P., Rajasthan, Gujarat, Haryana, Delhi, Madhya Pradesh and Maharashtra and has yield potential of 480-500 q/ha.

SHYAMAL (ARBH-201)

ARBH-201 has been developed through heterosis breeding at Ankur Seeds Private Limited, Nagpur and identified for release through AICRP-VC in 1993. Plants are semi-spreading to spreading with vigorous growth; fruits are purple and oblong with milky-white flesh having shelf-life of 4-5 days and good transport quality. It is tolerant to sucking borer pest and diseases. First picking starts after 100-110 days of transplanting and continues up to 240-270 days. This variety has been recommended for cultivation in Punjab, Bihar, U.P., Andhra Pradesh, Chhattisgarh, Orissa, Haryana, Rajasthan, Gujarat and Maharashtra and has yield potential of 700-750 q/ha.

MHB-39

MHB-39 has been developed through heterosis breesing (MBA-149 ×

MBA-1285) at MAHYCO Seeds, Jalna and identified for release through AICRP-VC in 1993. Plants are erect, compact and non-spiny. Fruits oval, shining, redish purple and born in bunches of 4 to 6 and non-spiny calyx. Being early bearing it takes about 60-65 days for first picking which continues up to 125-130 days. It has yield potential of 350-400 q/ha.

NARENDRA HYBRID BRINJAL-2 (NDBH-6)

NDBH-6 has been developed at NDUA&T, Faizabad and identified for release through AICRP-VC in 1995. Plants are non-spiny, semi-erect with green stem and leaves with rough surface and purpled veins. Fruits are attractive purple, born in clusters, long oval with soft flesh each weighing 150-900 g. Plants are moderately resistant to shoot and fruit borer, phomopsis blight and little leaf. This variety has been recommended for cultivation in Punjab, Bihar and U.P and has yield potential of 600-625 q/ha.

ANAND BRINJAL HYBRID-2 (ABH-2)

This hybrid has been developed through heterosis breeding (M-2 × M-31) at GAU, Anand and identified for release through AICRP (VC) in 1995. Fruits are light purple with green spiny calyx. This hybrid has been recommended for cultivation in Punjab, U.P., Bihar, Jharkhand, Rajasthan, Gujarat, Haryana and Delhi and has yield potential of 425-450 q/ha.

GULABI (SEL-4)

Sel-4 was developed at ANGRAU, Hyderabad and identified for release through AICRP-VC in 1995. Plants are erect with 5-6 primary branches, green stem with light purple shade and young leaves have green petiole. Fruits in cluster of 3-5, medium-long, light purple and highly suitable for long distance transportation. First picking starts 60-65 days after transplantation. This variety has been recommended for cultivation in Chhattisgarh, Orissa and A.P. and has yield potential of 325-350 q/ha.

AZAD B-2 (KS-224)

This variety has been developed through hybridization (Pusa Purple

Round × Arka Sheel) followed by pedigree selection at CSAU&T, Kanpur and identified for release through AICRP-VC in 1995. Plants are medium-tall (90 cm), compact with erect stem bearing 15-20 fruits. Fruits are round, purple with green shining calyx, solitary with medium thick pericarp and dull white firm flesh, average weight 135 g. First picking starts 55-60 days after transplanting and a fruiting duration of 140-150 days. This variety has been recommended for cultivation in Punjab, U.P. and Bihar and has yield potential of 300-325 q/ha.

PUSA HYBRID-9

Pusa Hybrid-9 has been developed at IARI, New Delhi and identified for release through AICRP-VC in 1997. Plants have green, non-spiny, strong erect branches with light pigmentation on younger leaves. Fruits are oval round, glossy, attractive, dark purple, peduncle partially pigmented with average weight of 300 g. It takes 90 days from sowing to first pickings and further pickings continue up to December. This hybrid has been recommended for the cultivation in Gujarat and Maharashtra and has yield potential of 675-700 q/ha.

KRISHNA (PHULE HYBRID-2)

This hybrid has been developed through heterosis breeding at MPKV, Rahurui and identified by AICRP-VC in 1997. Fruits are egg-shaped, purple with white stripes and calyx spiny. Plants are tolerant to fruit borer and little leaf disease. This hybrid has yield potential of 450-470 q/ha in 175 days of crop duration.

KS-331

This variety has been developed at CSAUA&T, Kanpur and identified for release through AICRP-VC in 1998. Plants are upright (erect stem), medium-tall (72 cm) with 3-5 branches and light green leaves bearing ~16 fruits. Fruits are long (15-19 cm), purple with average fruit weight of 218 g. This variety has been recommended for cultivation in Punjab, U.P. and Bihar, Chhattisgarh, Orissa and A.P. and has yield potential of 400-425 q/ha.

JB-15

This variety has been developed at JNKV, Jabalpur and identified for release through AICRP-VC in 1998. Plants are upright (erect stem), medium tall (83.9 cm), with 7-9 branches and green leaves bearing ~15 fruits. Fruits are long (~15 cm), violet purple with 270 g of average fruit weight. This variety has been recommended for cultivation in J&K, H.P. and Uttaranchal and has yield potential of 300-325 q/ha.

GREEN LONG

This variety has been developed at RAU, Sabour and identified for release through AICRP-VC in 1998. Plants are upright (erect stem), medium-tall (73 cm) with 8-9 branches and green leaves bearing ~8 fruits/plant. Fruits are long, green with 135g of average fruit weight. This variety has been recommended for cultivation in Punjab, U.P. and Bihar and has yield potential of 410-425 q/ha.

JAWAHAR BRINJAL-64 (JB-64-1-2)

This early variety has been developed through selection from JBP-2 at JNKV, Jabalpur and identified for release through AICRP-VC in 1998. The plants are spreading in nature and have 10-12 branches. The plant height ranges from 65-70 cm. Leaves are light green and flowers are purple. The fruits are ready for harvest within 130 days of seed sowing. Fruits are attractive purple, small, round, shining and each weighing 60-65 g. Seed to pulp ratio is ~1:27. This variety has been recommended for cultivation in Madhya Pradesh and Chhattisgarh and Gujarat and has yield potential of 320-350 q/ha.

NARENDRA BRINJAL-2 (NDB-28-2)

This variety has been derived from a cross (NDB-25 × NDB-2) at NDUA&T, Faizabad and identified for release through AICRP-VC in 2001. Fruits are long (20-25 cm), shining purple with partial pigmentation on peduncle, 100 g average fruit weight, soft texture and less-seeded. First picking starts at 70-75 days after transplanting. This variety has been recommended for cultivation in Punjab, U.P., Bihar and Jharkhand and has yield potential of 375-400 q/ha.

PUSA SHYAMLA (DBL-21)

This variety has been developed at IARI, New Delhi and identified for release through AICRP-VC in 2004. Plants are non-spiny with erect branches having light purple pigmentation on younger leaves. Fruits are long, glossy, attractive, dark purple with 80-90 g fruit weight, maturity period (days to first fruit harvest) is about 55 days from transplanting. This variety has been recommended for cultivation in Punjab, U.P., Bihar and Jharkhand and has yield potential of 390-410 q/ha.

ANAND BRINJAL SMALL ROUND-2 (ABSR-2)

This variety has been developed at AAU, Anand and identified for release through AICRP-VC in 2004. Fruits are small round, light pinkish, shiny with non-spiny calyx. This variety has been recommended for cultivation in Madhya Pradesh and Maharashtra and has yield potential of 410-425 q/ha.

AZAD B-4 (KS-235)

This variety has been developed at CSAUA&T, Kanpur and identified for release through AICRP-VC in 2004. Fruits are round light purple with green calyx. This variety has been recommended for cultivation in Punjab, U.P., Bihar, Jharkhand, Chhatishgarh, Orissa, A.P., Madhya Pradesh and Maharashtra and has yield potential of 380-400 q/ha.

BB-64

This variety has been developed at OUA&T, Bhubaneshwar, and identified for release through AICRP-VC in 2004. Plants are of medium height. Fruits are dazzling purple in colour, long, medium in size. Fruits mature in 95-100 days after sowing. Plants are resistant to bacterial wilt. This variety has been recommended for cultivation in Punjab, U.P., Bihar, Jharkhand, Chhattishgarh, Orissa, A.P., Madhya Pradesh, Maharashtra, Karnataka, Tamil Nadu and Kerala and has yield potential of 350-375 q/ha.

KASHI GANESH (IVBHL-54)

This hybrid has been developed at IIVR, Varanasi and identified for release through AICRP-VC in 2004. This is an excellent long fruited and prolific bearing hybrid having semi upright plant habit with plant height of 90-100 cm, light green stem and leaves. First flowering starts in 35-40 days after transplantation. Fruits are light purple, soft textured, long with average length and diameter of 13 cm and 3 cm, respectively. Fruit picking starts in 65-70 days after transplanting. This hybrid has been recommended for cultivation in Punjab, U.P., Bihar, and Jharkhand and has yield potential of 800-820 q/ha.

HYBRID ARBH-786

This variety has been developed through heterosis breeding (IC-470155 × IC-470154) at Ankur Seeds Private Limited, Nagpur and identified for release through AICRP-VC in 2008. It has semi-spreading growth habit with profuse branching. The calyx is green and non-spiny. First picking starts at 85-95 days after transplantation. Its fruits are purple, long, medium size each weighing 60-70 g. It is suitable for both early and late crop. This variety has been recommended for cultivation in Uttar Pradesh, Bihar and Jharkhand and has yield potential of 800 q/ha.

SWARNA ABHILAMB (HABL-1)

This variety has been developed through pure line selection from a germplasm collected from Dubri District of Assam at HARP, Ranchi and identified for release through AICRP-VC in 2006. Plants are tall (>100 cm), upright with intermediate spreading branches. The calyx is medium thorny and dark purple. It takes 55-60 days to flowering and fruits become ready for first harvest in 65-75 days after transplanting. Fruits are purple, 30-35 cm long and 80-100 g fruit weight and resistant to bacterial wilt. It is suitable for both winter and summer season cultivation. This variety has been recommended for cultivation in Uttarakhand, Himanchal Pradesh and Jammu and Kashmir and has yield potential of 500-600 q/ha.

CHILLI

Capsicum annuum
Family: Solanaceae
Chromosome no.: 2n=24

Chilies are green or dried ripe fruits of pungent forms of *Capsicum annuum* L. or *C. fruitescence*. India is one of the leading producer of this economically important vegetable. China, Korea, Nigeria, Thailand, Turkey, Kenya, Sudan, Uganda, Japan, Ethiopia, Indonesia, Pakistan and Mexico are the other chilli growing countries. The produce is mostly consumed in India and only a small quantity (1 to 2%) is exported to various countries. In India, the major chilli growing states are Andhra Pradesh, Maharashtra, Karnataka and Tamil Nadu, which together constitute ~75% of the total area. Andhra Pradesh ranks first in dry chilli production followed by Tamil Nadu, Maharashtra, Orissa and Karnataka. Chillies are considered to be native of South America, possibly Peru. Mexico is the secondary centre of origin and diversity. In India, chilli was introduced by Portuguese probably in late 16th Century. There are five species under cultivation. The two species, *i.e.*, *C. pendulum* (syn *baccatum*) and *C. pubescens* are restricted to South and Central America, while *C. chinense* is cultivated in Africa. The other two species, *C. annuum* and *C. fruitescens* are commercially cultivated species.

Green chillies are a rich source of vitamins A and C. Each 100 g of green fruits contain moisture (85.7 g), proteins (2.9 g), Fat (0.6 g), minerals (1.0 g), Fibre (6.8 g) and carbohydrates (3.0 g). Chillies are used as vegetables, spices, condiments, sauces, sambar, rasam and pickles. Dry chillies are used as curry powder. In addition, chilli extracts are used in the production of ginger beer and other beverages. Capsaicin, an alkaloid is extracted from chillies and is used in medicine. *C. fruitescens* is used in medicine as carminatives. The pungency in chilli is mainly due to the alkaloid Capsaicinoids, which occur in core/septa walls and in placenta. The bright red colour at ripening stage is due to carotenoid pigment Capsanthin. In food and beverage industries, chilli is being used in the form of oleoresin, which permits better distribution of colour, and flavour in food. Chilli is grouped as often cross-pollinated crop, the anthesis of flowers takes early in morning. Maintenance breeding is very important in chilli to maintain the purity, this is usually achieved by selfing at bud stage or caging the plant individually or collectively with nets.

Chilli can be cultivated both in tropical and sub-tropical regions. Chilli plants require temperature of 20-25°C fluctuations in temperature and water stress during flowering affect the bud opening and fruit drops. In India, chilli is commercially grown during monsoon or kharif season (June-October) as unirrigated crop or with little irrigation. Well-drained sandy loam soil is preferred for chilli cultivation. In southern parts of the country, seeds are sown in the nursery from mid of May to July. In Punjab and Rajasthan, it is shown in April-May. In hilly areas, it can be sown in March-April. About 500 g seeds are required to raise seedlings for hectare land. 4 week old seedlings are generally used for transplanting. Fruits are harvested either as green or red ripe by hand picking. Frequent pickings (about 6) at 10-12 days interval are required for green fruits. Damping off, anthracnose, leaf curl, and chilli mosaic are important diseases, whereas thrips, aphids and mites are important insects causing heavy yield loss.

Varieties

In India, ICAR institutes and SAUs are working for the improvement of chilli. The major research work on chilli is going on at IIVR (Varanasi), PAU (Ludhiana), ARS (Lam), IIHR (Bangalore), AAU (Anand), UAS (Dharwad) and CSAUA&T (Kanpur). Generally, selection and heterosis breeding have been used for genetic improvement and several varieties/hybrids have been released and recommended for cultivation under different agro-climatic zones as described below (for photographs see plate 4).

HOT PORTUGAL (CA-960)

This variety has been developed at ARS, Lam and notified by CVRC Notification no. 440 dated 21.08.1975. Plants are tall with green foliage. Fruits have deep red and thick pericarp. Fruit measures 8.2 cm in length with capsaicin content of 0.81 mg/g of fruit. Initially, the pericarp of fruit is light green turning deep red on maturity with smooth surface. Top of the fruit is blunt. Fruit has mild pungency. Seed content is 38%. It has yield (dry chilli) potential of 12 q/ha under rainfed and 60-80 q/ha under irrigated conditions.

ANDHRA JYOTI (G-5)

This variety has been developed through hybridization (G-2 x Bihar Variety) followed by pedigree selection at ARS, Lam and notified by CVRC Notification no. 1004(E) dated 23.03.1978. Plants are tall and dense. Fruits of this cultivar are short and gundu type. Fruits are bright red measuring 5.1 cm in length and 6.3 cm in girth. The capsaicin content is 0.65 mg/g of fruit. This variety has been recommendede for cultivation in J&K, H.P., Uttaranchal, Punjab, U.P., Bihar, Chhattisgarh, Orissa, A.P., Rajstahan, Gujrat, Harayana, Delhi, M.P., Mharastra, Karnataka, Tamilnadu and Kerala and has yield (dry chilli) potential of 20-30 q/ha.

SINDHUR

Sindhur is a dual-purpose chilli selected from Hot Portugal (CA-960) developed at ARS, Lam and notified by CVRC Notification no. 13 dated 19.12.1978. Plants are tall with green foliage. Fruits have deep red and thick pericarp. Fruit measures 8.2 cm in length with capsaicin content of 0.81 mg/g of fruit. Pericarp of fruit is light green turning deep red on maturity with smooth surface. Top of fruit is blunt. Fruits are mild pungent. Seed content is 38 %. This variety has yield potential of 12-14 q/ha (dry chilli) under rainfed and 60 q/ha under irrigated conditions.

PANT C-1

This variety has been developed through generation advancement and selection from a cross (NP-46-A × Local cultivar) at GBPUA&T, Pantnagar and notified by CVRC Notification no. 19(E) dated 14.01.1982. Plants are of short stature with more primary branches. Fruits are erect, small-sized, highly pungent, 5.5 cm long, light green and turning light red at maturity. First fruiting commences in two months and picking commences 100 days after transplanting. It is tolerant to mosaic and leaf curl viruses and has yield (dry chilli) potential of 12-15 q/ha.

CO-1

This variety is a selection from Sattur Samba (CA-P-247) has been developed at TNAU, Coimbatore and notified by CVRC Notification no. 19(E) dated 14.01.1982. Plants are erect, medium tall and compact with medium branching. Fruits are green when unripe and bright shiny red on ripening. Fruits are 6-6.5 cm long with sharp tip and bulge shoulders. The seed content is 55 per cent. The variety has high capsaicin content (0.72 mg/g) and has yield potential of 21 q/ha (dry chilli) in 210 days of crop duration.

DH-76-6

This variety has been developed thropugh hybridization (Puri Red x Byadgi Local) followed by pedigree selection at UAS, Dharwad and notified by CVRC Notification no. 19(E) dated 14.01.1982. Plant height is 70-75 cm, highly branched with bushy growth habit. Stem is light green. Leaves are smaller than Byadgi variety. It has 25-30 fruits/ plant. Fruits are thin (~2 mm pericarp), pungent, long, light red with wrinkles. Seeds are light yellow with seed recovery of 60 %. Seed yield is 1.2 q/ha. This variety has yield potential of 20 q/ha (dry chilli) in 180 days of crop duration.

PUSA JWALA

This 'Samba' type variety has been developed through hybridization (NP-46-A × Puri Red) followed by predigree selection at IARI, New Delhi and notified by CVRC Notification no. 499(E) dated 08.07.1983. Plants are dwarf, bushy, light green with broad leaves. Fruits are 9-10 cm long and thin, light green turning light red at maturity. Dry fruits are wrinkled and liable to break during packing and transport. The fruit contain capsaicin 0.48 mg/g of fruit, thus are highly pungent. It is tolerant to thrips, mites and aphids and has yield (green chilli) potential of 75-90 q/ha.

CO-2

This variety has been developed at TNAU, Coimbatore and notified by CVRC Notification no. 596(E) dated 13.08.1984. It is a selection form Nambiyar local Gundu type (CA-P-63) of Periyar district of Tamil Nadu. The stem is angular, semi-dwarf and less-spreading. Fruits are oblong, thick and bright red. Seed content is high (60%). The capsaicin content of dry pod is 0.56%. It is suitable both as green pods and red ripe pods. It has yield (green chilli) potential of 110-125 q/ha in a crop duration of 210 days.

CHANCHAL

This variety has been developed at CSAUA&T, Kanpur and notified by CVRC Notification no. 295(E) dated 09.04.1985. Plants are medium tall with perennial habit. Foliage is dark green. Fruits are 25-30 cm long, 6-7 cm thick, erect, highly pungent, dark green when fresh and shining red on maturity. Plants are tolerant to leaf curl disease. It is suitable for kitchen garden and has yield (dry chilli) potential of 15-20 q/ha.

K-2

This variety has been developed through hybridization (K-1 x Sathur Samba) followed by pedigree selection at TNAU, Kovilpatti and notified by CVRC Notification no. 540(E) dated 24.07.1985. Plants are tall and semi-spreading. Fruits are long (7.3 cm) with 0.19 mm pericarp thickness. This variety has been recommended for cultivation in J&K, H.P, Uttaranchal, Chhattisgarh, Orissa, A.P., Rajasthan, Gujarat, Haryana and Delhi and has yield (dry chilli) potential of 9-12 q/ha.

MDU-1

This variety has been developed through mutation breeding at TNAU, Coimbatore and notified by CVRC Notification no. 540(E) dated 24.07.1985. It is an induced mutant of K-1 using 30 kr of gamma rays. It possesses broad and dark green leaves, compact plant type and determinate growth habit with less number of secondary branches. Fruits are borne in clusters of 4 to 9 at nodes as against single fruit borne at nodes as in K-1 or K-2 varieties. Fruits are long with dark shiny red colour. This variety has yield (dry chilli) potential of 18-20 q/ha in a crop duration of 205-215 days.

K-1

This variety is a pure line selection from Local Sattur Samba, has been developed at TNAU, Kovilpatti and notified by CVRC Notification no. 832(E) dated 18.11.1985. The plants are tall and spreading. The crop duration is 210 days. The ripe fruits are red, long and contain high capsaicin. It is suitable for rainfed cultivation in southern district of Tamil Nadu and has yield (dry chilli) potential of 18-20 q/ha in 120 days of crop duration.

BHASKER (LCA-235)

This is an early maturing variety developed through hybridization (G-4 × Yellow Anther Mutant) followed by pedigree selection at ARS, Lam and notified by CVRC Notification no. 867(E) dated 26.11.1986. Plants are compact, spreading with dark green leaves and yellow anthers. Fruits are thin, medium-long, thick red with red flesh and 45% seed content, pungent with high oleoresin, suitable for export, good shelf-life and first picking starts at about 130 days after sowing. Plants are tolerant to mite, thrips and aphids. This variety has been recommended for cultivation in Punjab, U.P., Bihar, Chhattisgarh, Orissa, A.P., Rajasthan, Gujarat, Haryana, Delhi, Karnataka, Tamil Nadu and Kerala and has yield (dry chilli) potential of 15-17 q/ha.

PUNJAB LAL

This variety has been developed through hybridization (Perennial × Long Red) followed by pedigree selection at PAU, Ludhiana and notified by CVRC Notification no. 165(E) dated 06.03.1987. Plants are dwarf, bushy with dark green foliage. Fruits are erect, medium-sized (4.25 × 0.79 cm), dark green and turning dark red at maturity. Fruits are rich in capsaicin (0.7%), oleoresin, dry matter and red pigment. Plants are resistant to CMV and leaf curl viruss and moderately resistant to fruit rot and die back diseases. It has yield (green chilli) potential of 100-120 q/ha.

JAWAHAR MIRCH-218 (J-218)

This variety has been developed through hybridization (Kali Peeth × Pusa Jwala) followed by pedigree selection at JNKV, Jabalpur and notified by CVRC Notification no. 915(E) dated 06.11.1989. Plants are dwarf (50-55cm), spreading with light green foliage. Fruits are long, thin, bright green; good transport quality. It is tolerant to leaf curl and fruit rot; first picking starts 100-105 days after sowing. This variety has been recommended for cultivation in Punjab, U.P., Bihar, Chhattisgarh, Orissa, A.P., Rajasthan, Gujarat, Haryana and Delhi, Karnataka, Tamil Nadu and Kerala and has yield potential of 90-100 q/ha (green chillies) and 20-30 q/ha (dry chilli).

MUSALBADI

This variety has been developed at MPKV, Rahuri and notified by CVRC Notification no. 527(E) dated 16.08.1991. Musalbadi is derived from a local collection. Plants are tall, dark green, spreading foliage with black patches. Fruits are smooth, medium-long, dark green and dark red at maturity, which are consumed as green as well in dry. Plants are tolerant to die back and powdery mildew diseases. This variety has been recommended for cultivation in Punjab, U.P., Bihar, eastern M.P., Orissa, A.P., Rajasthan, Gujarat, Haryana and Delhi, Karnataka, Tamil Nadu and Kerala and has yield (green chilli) potential of 110-125 q/ha in 180 days of crop duration.

PKM-1

This variety has been developed at TNAU, Coimbatore and notified by CVRC Notification no. 793(E) dated 22.11.1991. PKM-1, is a derivative of the cross CAN-1797 × CO-1. Plants are dwarf (70 cm) and fruits are green and dark red. It is suitable for cultivation under irrigated conditions. The number of fruits/plant is 204 with 85 seeds/ fruit. This variety has yield potential of 30-40 q/ha in 180 days of crop duration.

LCA-305 (LAM-305)

This variety has been developed at ARS, Lam and notified by CVRC Notification no. 408(E) dated 04.05.1995. Plants are bushy and fruits are light green while tress but become bright red after drying. It is tolerant to leaf curl virus and has yield (dry chilli) potential of 20-25 q/ha.

JAYANTI (AKC-86-39)

This variety has been developed at PDKV, Akola and notified by CVRC Notification no. 115(E) dated 10.02.1996. Fruits are green while fresh and turn red on ripening. Plant height is medium to tall (60-75 cm) with light green foliage. This variety has yield (dry chilli) potential of 20-30 q/ha.

PUNJAB GUCHHEDAR

This variety has been developed at PAU, Ludhiana and notified by CVRC Notification no. 115(E) dated 10.02.1996. Fruits are small (4.68 cm), erect and borne in clusters. Fruits are deep red, suitable for drying and processing. This variety has yield (dry chilli) potential of 15-17 q/ha.

PUNJAB SURKH

This variety has been developed at PAU, Ludhiana and notified by CVRC Notification no. 115(E) dated 10.02.1996. Fruits are long (6.68 m) and deep red on maturity. It is very good for powder preparation. It is tolerant to fruit rot and resistant to viral diseases and has yield (dry chilli) potential of 20-25 q/ha.

JAWAHAR-283

This variety has been developed through selection from a local genotype at JNKV, Jabalpur and notified by CVRC Notification no. 1052(E) dated 26.10.1999. Plants are dwarf (40-45 cm) with bushy habit. It produces 7-9 primary branches with dense foliage cover. Fruits are straight, firm, medium in length, thin (2.5-3.5 cm girth in the middle) and dark green. Ripe fruits are bright red and pungent. It has yield (green chilli) potential of 80-100 q/ha.

KASHMIR LONG-1

This variety has been developed at SKUA&T, Srinagar and notified by CVRC Notification no. 1135(E) dated 15.11.2001. Plants are medium tall with spreading habit, ovate green leaves, round to angular stem with purple pigmentation on nodes. Bright red pendent, long fruits with yellow seeds. It is moderately resistant to Fusarium wilt and has yield (green chilli) potential of 160-180 q/ha.

CO-3

This variety is a selection from Sri Lankan collection, has been developed at TNAU, Coimbatore and notified by CVRC Notification no. 1135(E) dated 15.11.2001. Plants are dwarf and compact. It is

Samba type chilli and suitable for very close spacing planting (30 × 15 cm). It is suitable for dry as well as green chilli production. It has very low stalk weight in comparison with pod weight. Due to high oleoresin and capsaicin contents it is suitable for export. It has yield (green fruits) potential of 350-375 q/ha.

UTKAL AVA (BC-14-2)

This variety has been developed at OUA&T, Bhubaneswar and notified by CVRC Notification no. 642 (E) dated 31.05.2004. Fruits are upward at plant, deep red at ripening with good pungency suitable both for green and dry chilli purpose. Fruits can be harvested in 125-130 days after sowing. Fruit length is 6.2 cm and girth is 3.7 cm. It is resistant to bacterial wilt disease and has yield potential of 120-140 q/ha (green chilli) and 35-50 q/ha (dry chilli).

ARKA SUPHAL (PMR-57-88-K)

This variety has been developed through hybridization (Pant C-1 × IIHR-517A) followed by pedigree selection at IIHR, Bangalore and notified by CVRC Notification no. 597(E) dated 25.04.2006. Plants tall and erect with green, smooth, medium long (6-7 cm × 1 cm), pendent fruits. It is tolerant to powdery mildew. It is suitable for cultivation in irrigated and rainfed condition areas and has yield potential of 230-250 q/ha (green fruits) and 30-40 q/ha (dry chilli) in 180 days of crop duration.

GUJARAT VEGETABLE CHILLI-101 (ACS-2000-02)

This variety has been developed through selection from a predominant local variety S-49 at AAU, Anand and notified by CVRC Notification no. 597(E) dated 25.04.2006. Fruits are medium pungent, shining green, long, compact and straight. Fruits have longer shelf-life and high ascorbic acid content. This variety has been recommended for cultivation in Madhya Pradesh and Maharashtra and has yield potential of 140-150 q/ha.

KASHI ANMOL (KA-2)

This variety is an introduction from Sri Lanka, recommended by IIVR, Varanasi and notified by CVRC Notification no. 597(E) dated

25.04.2006. Plants are determinate, bushy with nodal pigmentation on stems. Fruits pendant, attractive green, first picking can be taken 50 days after transplanting. It is suitable for green fruit production under chilli-wheat/chilli-potato cropping system. This variety has been recommended for cultivation in Punjab, U.P., Bihar and Jharkhand and has yield (green chilli) potential of 200-225 q/ha in a crop duration of 130-145 days.

PRASHANTH (LCA-334)

This early maturing variety has been developed at ARS, Lam and notified by CVRC Notification no. 597(E) dated 25.04.2006. Plants are compact, spreading with dark green leaves and yellow anthers. Fruits are medium-long, green when immature and glossy red on ripening, suitable for both green and dry purposes, very good for export having good shelf-life and transport quality. First picking starts about 130 days after sowing. Plants are tolerant to mites, thrips and aphids. This variety has yield (green chilli) potential of 150-170 q/ha.

ARKA LOHIT (SEL-1)

This variety has been developed through selection from a local collection (IHR-324) at IIHR, Bangalore and notified by CVRC Notification no. 2035(E) dated 28.11.2006. Fruits are dark green, smooth, straight, turn dark red at maturity, highly pungent (capsaicin 0.70%). This variety has been recommended for cultivation in Chhattisgarh, Orissa, A.P., M.P, Maharashtra, Karnataka, Tamil Nadu and Kerala and has yield potential of 140-150 q/ha (dry chilli) in 180 days of crop duration.

AZAD MIRCH-1

This variety has been developed at CSAUA&T, Kanpur and notified by CVRC Notification no. 597 (E) dated 25.04.2006. Plants are dwarf, upward fruiting type, branched with early maturing length medium fruits. Light green fruits are borne in clusters of 8 to 12. This variety has yield (dry chilli) potential of 50-60 q/ha.

KASHI SURKH (CCH-2)

This hybrid has been developed through heterosis breeding (MS line;

CCA-4261 × Pusa Jwala) at IIVR, Varanasi and notified by CVRC Notification no. 2035(E) dated 28.11.2006. Plants are semi-determinate (1-1.2 m), erect with pigmentation on stem. Fruits are light green, straight, 11-12 cm length, suitable for green as well as red fruit production. The first harvest starts after 55 days of transplanting. Its plants are tolerant to thrips, mites and viruses. This variety has been recommended for cultivation in West Bengal, Assam, Punjab, Uttar Pradesh, Jharkhand, Bihar, Chhattisgarh, Orissa, A.P., M.P. and Maharastra and has yield potential of 220-240 q/ha (green fruit) and 140-150 q/ha (dry chilli).

ANUGRAHA

This variety has been developed at KAU, Vellanikkara and notified by CVRC Notification no. 597 (E) dated 25.4.2006. Plants are indeterminate, solitary, linear and pendulous, resistant to bacterial wilt disease. Fruits are light green at immature stage and red on ripening. Average fruit length 12 cm and average fruit weight is 3.6 g. It has yield (green chilli) potential of 270-300 q/ha.

GUJARAT VEGETABLE CHILLI-111 (ACS-2000-03)

This variety has been developed through selection from the variety Jwala at Anand Agricultural University, Anand and notified by CVRC Notification no. 597 (E) dated 25.04.2006. Fruits are attractive, medium-long, straight, pungent and slightly wrinkled surface with good shining. This variety has been recommended for cultivation in middle Gujarat and has yield (green chilli) potential of 110-125 q/ha.

GUJARAT VEGETABLE CHILLI-121 (ACS-97-2)

This variety has been developed through hybridization (G-4 × S-49) followed by pedigree selection at Anand Agricultural University, Anand and notified by CVRC Notification no. 597 (E) dated 25.04.2006. Fruits are longer, thick, dark green with good shining and possess higher ascorbic acid content and better shelf-life. This variety has been recommended for cultivation in middle and south Gujarat and has yield (green chilli) potential of 100-120 q/ha.

RCH-1

This variety has been developed at RARS, Durgapura and notified by CVRC Notification no. 642 (E) dated 31.05.2004. Plant height is 95 cm. It takes ~50 days to flower and bear 80 fruits per plant. Picking starts 110 days after transplantaion and harvesting is completed in ~200 days. Fruits are long and bright red, 13.1 cm length and 3.4 cm girth. The variety is well suited to arid and semi arid conditions of Rajasthan and has yield potential of 20-30 q/ha of dry chilli.

BHAGYALAKASHMI (G-4)

This variety has been developed at RARS, Lam, Guntur and identified for release through AICRP-VC in 1975. G-4 has been derived through selection in Thohian chilli from Sri Lanka. Leaves are dark green. Fruits are olive green, trun dark red on ripening, long (7-8 cm), 3-35 cm girth with 38-40% seed content. Plants are fairly tolerant to diseases and pests. It has been recommended for cultivation in J&K, H.P., Uttaranchal, Punjab, U.P., Bihar, Chhattisgarh, Orissa, A.P., Rajasthan, Gujarat, Haryana, Delhi, M.P., Maharashtra, Karnataka, Tamil Nadu and Kerala and has yield potential of 80-100 q/ha under irrigated and 50-60 q/ha under rainfed conditions.

PHULE SURYAMUKHI (RHRC-CLUSTER ERECT CHILLI)

This variety has been developed through selection from a germplasm at MPKV, Rahuri identified for release through AICRP-VC in 2001. Plants spreading, fruits smooth and dark green, dark red at maturity, born in cluster of 3-5. Plants are tolerant to leaf curl. It has been recommended for cultivation in M.P. and Maharashtra and has yield potential of 28-30 q/ha of dry chilli in 190 days of crop duration.

PRAKASH (LCA-206-B)

LCA-206 has been developed through hybridization (G-3 × Huntaka) followed by pedigree selection at RARS, Lam, Guntur and identified for release through AICRP-VC in 1990. Plants are tall with narrow light green leaves. Fruits are long, thin, bright red at maturity with red flesh. Picking starts at about 135 days after transplantation. Plants are moderately tolerant to sucking pests, bacterial leaf spot and fruit rot. It is suitable for rainfed conditions. This variety has been recommended for cultivation in Chhattisgarh, Orissa, A.P., M.P., Maharashtra, Karnataka, Tamil Nadu, Kerala and has yield potential of 80-100 q/ha.

HYBRID CHILLI (ARCH-236)

ARCH-236 has been developed at Ankur Seeds Private Limited, Nagpur and idenfied for release through AICRP-VC in 1997. Plants have profuse branching, bearing elongated, undulated, pointed and green fruits. It has better shelf-life, high capsicin content and suitable for fresh market. This hybrid has been recommended for cultivation in states of Punjab, Bihar and U.P. and has yield potential of 425-450 q/ha crop duration of 210-140 days.

SUNGRO-86-235

This variety has been developed by Sungro Seeds and identified for release through AICRP-VC in 2001. The plants of this hybrid are umbrella shaped. Fruit are dark green which turns to medium red at maturity. Fruits are pendant, long (10-11 cm) and 0.8 cm in diameter and highly pungent. This hybrid has been recommended for cultivation in Punjab, U.P., Bihar, Jharkhand, Madhya Pradesh and Maharashtra and has yield potential of 70-80 q/ha.

LCA-353

This variety has been developed through hybridization (LCA-305 × Sel.-116) followed by pedigree selection at RARS, Lam and identified for release through AICRP-VC in 2007. Plants are of medium height, fruits are slender and shiny red with 9-10% oleoresin. It is suitable for green and dry chilli production as well as for processing and export. This hybrid has been recommended for cultivation in Chattisgarh, Orissa, Andhra Pradesh, Kerala, Punjab, U.P. and Bihar and has yield (dry chilli) potential of 75 q/ha.

SWEET PEPPER

Capsicum annuum
Family: Solanaceae
chromosome no.: 2n=24

In India, cultivation of sweet pepper (syn. bell pepper, capsicum) is recent, thus as compared to other Solanaceous vegetables, sweet pepper is cultivated at relatively small scale. In the recent past, however, its cultivation has spread in the northern plains and it has become a high value crop. Commercial hybrid seeds are being mainly produced in temperate regions and southern parts of India through hand emasculation and pollination. Despite high cost of hybrid seeds, there has been increase in the cultivation of hybrids. This is because under optimum crop production and protection management, crop raised from the seeds of F_1 hybrid has several advantages like better yield, adaptability, uniformity and reactions to certain stresses in comparison to crop raised from the seeds of improved pure line or population.

Capsicum is mostly consumed as cooked vegetable in India, whereas in Europe it is consumed as raw. The genus *Capsicum* originated in the new world tropics and sub tropics. The major countries producing bell pepper are China, Indonesia, Sri Lanka, Pakistan, Turkey, Korea, Hungary, Spain, Bulgaria, Romania, Italy, Yugoslavia, Nigeria, Ghana, USA, Central America, Peru, Tunisia, Mexico and Argentina. In India, the production of capsicum is restricted around the periphery of chilli. *Capsicum* is generally cultivated around cities like Bangalore, Belgaum, Mysore (Karnataka), Nilgiris (Tamil Nadu), Pune, Thane (Maharastra), Ranchi (Jharkhand), Darjeeling (West Bengal), and in Himachal Pradesh, Jammu & Kashmir and Hilly trace of northern plains.

Anthesis of flower takes early in the morning and continue upto 8 AM. The stigma is receptive from 24 hours before and 48 hours after anthesis. Pollens remain viable on the day of anthesis, whereas it is fertile before 12 hours of anthesis. *Capsicum* varieties can be maintained through manual selfing or caging the plant individually or collectively with nets of 40 mesh size. Commercial seed production can be done at 400-500 m isolation distance.

The major objectives of capsicum breeding are attractive fruit colour at immature stage, bell shape and bigger size of the fruits. The diseases, *i.e.* virus complex, anthracnose and insects (aphid, mite and thrips) are in the priority for improvement of sweet papper. Selection, pedigree selection and backcross pedigree selection are important breeding methods followed to develop open pollinated varieties. Heterosis breeding using hand emasculation and pollination is used to develop the commercial F_1 hybrids. Cost of hybrid seed production could be minimized by adopting *cms* lines.

Varieties

In India, ICAR Institutes and SAUs are working for the genetic improvement of capsicum. The major research work on sweet pepper is under progress at IARI Regional Station (Katrain), HPKV (Palampur), and DARL (Pithoragarh). Generally, selection, hybridization and heterosis breeding have been used for improvement of sweet pepper and several varieties/hybrids have been released and recommended for cultivation under different agro-climatic zones as described below (for photographs see plate 3).

CALIFORNIA WONDER

This variety is an introduction from USA and notified by CVRC Notification no. 440 dated 21.08.1975. Plants are vigorous, upright and prolific. Fruits are smooth, 3-4 lobed with medium thick flesh. It has yield potential of 125-150 q/ha in 125-140 days of crop duration.

CHINESE GIANT

This variety is an introduction from USA and notified by CVRC Notification no. 440 dated 21.08.1975. Plants are vigorous and prolific bearers. Fruits are smooth, 3-4 lobed, medium thick with dark green skin colour. It has yield potential of 130-150 q/ha in 125-150 days of crop duration.

BULL NOSE

This variety is an introduction and notified by CVRC Notification no.

13 dated 19.12.1978. Plants are 50-60 cm tall, erect and sturdy. Foliage is dark green. Fruits are 8-10 cm long, thick walled 3 lobed, and deep green. Flesh is thick with non-pungent. Fruit becomes scarlet red on maturity. Flowering takes place 80-85 days after transplanting. It has yield potential of 100-120 q/ha.

ARKA BASANT

This variety has been developed at IIHR, Bangalore and notified by CVRC Notification no. 867 (E) dated 26.11.1986. This variety is an improvement over a Hungarian variety Soroksari. It is creamish white, conical, erect fruited variety with thick flesh and prolific bearing. It is having good aroma and crisp texture. The variety is suitable for both *kharif* and *rabi* seasons as well as under protected conditions. It has got good export value of the fruits and has yield potential of 150-200 q/ha in 125 to 140 days of crop duration.

ARKA GAURAV

This variety is a pure line selection from Golden Calwonder from USA, developed at IIHR, Bangalore and notified by CVRC Notification no. 867(E) dated 26.11.1986. Plant is indeterminate with dark green fruits. Fruits are blocky, 3-4 lobed, thick fleshed and medium-sized (70-80 g). Fruits turn to orange yellow on maturity. It is good for both *kharif* and *rabi* seasons. Foliage covers the fruits to avoid sunscald. It is tolerant to bacterial wilt disease and has yield potential of 180-250 q/ha in 125 to 150 days of crop duration.

ARKA MOHINI

This variety is a selection from a variety Titan from USA, developed at IIHR, Bangalore and notified by CVRC Notification no. 867(E) dated 26.11.1986. Plants are determinate with medium to large blocky fruits. Fruits have dark green surface, flesh is thick 3-4 lobed and become red on ripening. This variety is suitable for both *kharif* and *rabi* seasons. Foliage covers the fruits avoiding the sunscald. Tipping of initial flowers is advocated to get higher yield. It has yield potential of 250-275 q/ha in 120 days of crop duration.

SOLAN HYBRID-1

This variety has been developed at HPKV, Solan and notified by CVRC Notification no. 115 (E) dated 10.02.1996. Fruits are 3-4 lobed and bell shape. This variety has been recommended for cultivation in H.P. and Uttaranchal and has yield potential of 300-350 q/ha.

PUSA DEEPTI (PAPRIKA TYPE; KT-P-119)

This variety has been developed at IARI, Regional Station, Katrain and notified by CVRC Notification no. 98(E) dated 8.02.1997. Plants are upright, spreading with profused branching. Fruits are smooth, erect, conical, light green with thick flesh, 9-11 cm long and 3-5 cm in diameter. The fully ripe fruits are ready for picking in 70-80 days after transplanting. Each plant on an average bears 55-59 fruits.. It can be grown successfully in spring-summer (February-June) in hills and northern plains of India. It is tolerant to bacterial leaf spot and anthracnose under field conditions and has yield potential of 250-350 q/ha.

NISHAT-1 (SEL-2)

This variety has been developed at SKUAS&T, Srinagar and notified by CVRC Notification no. 1135 (E) dated 15.11.2001. Plants are indeterminate, medium (48 cm) with dark green foliage, green stem with purple strips. Fruits are dark green, turn yellow at physiological maturity, 3-4 lobed, non-pungent and bell-shaped. First picking starts at 45-50 days after transplanting. This variety has been recommended for cultivation in J&K, H.P. and Uttaranchal and has yield potential of 200-225 q/ha.

DARL-202

This hybrid has been developed at DARL, Pithoragarh and identified for release through AICRP-VC in 2002. Plants are early maturing with vigorous growth. Fruits are three to four lobed. Each plant bears 18-20 fruits. It is recommeneded for cultivation in Punjab, U.P., Bihar, Jharkhand, J&K, H.P. and Uttaranchal and has yield potential of 340-370 q/ha.

KTCPH-3

This variety has been developed at IARI, Regional Station, Katrain and identified for release through AICRP-VC in 2005. Plants erect, medium and bushy. First picking in 60-70 days after transplanting. Fruits are yellowish green, conical, turns dark red at maturity; 9-11 cm long and very early in fruiting. It is recommeneded for cultivation in J&K, H.P., Uttaranchal, Rajasthan, Gujarat, Haryana, Delhi, M.P. and Maharashtra and has yield potential of 225-250 q/ha.

Chapter 3

Legume Vegetables

VEGETABLE PEA

Pisum sativum
Family: Papillionaceae
Chromosome no.: 2n=14

Pea is a nutritious leguminous vegetable grown throughout the world. In India, it is grown as winter vegetable in the plains of North India and as a summer vegetable in the hills. Central Asia is considered to be the centre of origin of pea. The genus *Pisum* includes 6-7 species, mostly found in the Mediterranean area and West Asia, out of which only *P. sativum* is mainly cultivated. Pea flowers are solitary, up to 3 flowers per raceme, bracts very small, calyx oblique, lobes unequal, corolla white, pink or purple, keel short, incurved, obtuse, stamens diadelphous, filamentous broad, anthers uniform, style falcate, flattened, bearded on inner surface, stigma minute terminal. The extent of natural outcrossing has been estimated to be ~1 %.

Pea can be grown in well-drained loamy soil. The most favourable range of pH is 5.5-6.0. It can be grown well at the temperature 10-18°C. Although the optimum temperature for germination is ~22°C but seeds can also be germinated at 5°C. Pea varieties are grouped as early, mid and late maturity. Therefore, sowing can be done from beginning of October to mid November in the plains of northern India. Varieties of early maturity group generally suffer from wilt, whereas, late-maturing varieties have mildew problem.

Peas are diploid and self-pollinated crop. Hybridization involving cultivars, land races and primitive forms, followed by pedigree, single seed descent, bulk or back cross methods are traditionally used for its genetic improvement.

Varieties

In India, ICAR Institutes and SAUs are working for the improvement of pea. The major research work on pea is going on at IIVR (Varanasi), CSAUA&T (Kanpur) and GBPUA&T (Pantnagar). The main objectives of pea breeding are attractive seeds and resistance against biotic and abiotic stresses. Several varieties of different maturity groups have been developed as detailed below (for photographs see plate 5).

BONNEVILLE

This medium maturity group variety has been developed through an introduction from USA at IARI, New Delhi and notified by CVRC Notification nos. 716, 361(E), 440 dated 20.02.1970, 30.06.1973, 21.08.1975, respectively. Plants are medium-tall, flower in 55-60 days after seed sowing with two pods per peduncle. Pods are light green, straight, ~9 cm long, having 6-7 grains per pod with 45% of shelling. Seeds are green and wrinkled. This variety has been recommended for cultivation throughout India and has yield potential of 65-70 q/ha.

ARKEL

This early variety is an introduction from U.K. recommended by IARI, New Delhi and notified by CVRC Notification no. 1004 dated 23.03.1978. Plants are dwarf (35-45 cm), bear double pods at lower nodes and solitary at upper nodes. Pods are attractive, dark green, ~9 cm long, curved towards suture at distal end (sickle shaped), well filled with 7-8 seeds/pods with 40% shelling; suitable for fresh market and dehydration. Plants are highly susceptible to collar rot if sown under high temperature. This variety has been recommended for cultivation throughout India and has yield potential of 75-80 q/ha in 50-55 days of crop duration.

VL MATAR-3 (VL-3)

This medium maturity group variety has been developed at VPKAS, Almora and notified by CVRC Notification no. 1135(E) dated 01.12.1988. Plants are determinate with light green foliage, flowers white and bear two pods per peduncle. Pods are light green, straight with 6.8 cm of average length, first picking in 100 and 140 days after sowing in the plains and hills, respectively. Plants are moderately tolerant to powdery mildew and highly tolerant to cold at seedling stage. This variety has been recommended for the cultivation in J&K, H.P., Uttaranchal, Punjab, U.P., Bihar, Rajasthan, Gujarat, Haryana and Delhi and has yield potential of 100-105 q/ha.

VL AGETI MATAR-7 (VL-7)

This early maturity group variety has been developed at VPKAS, Almora and notified by CVRC Notification no. 408(E) dated 04.05.1995. Plants are determinate, dwarf with dark green foliage and white flowers. Pods are light green, slightly curved towards suture at distal end, medium (~8 cm), completely filled, two pods in a bunch, first picking in 120-125 days after sowing; pods contain 6-7 seeds, seeds are light green and wrinkled with 42% of shelling. Plants are more tolerant to *Alternaria* blight than Arkel. This variety has been recommended for cultivation in Punjab, U.P., Uttarakhand and Bihar and has yield potential of 230-250 q/ha green pod in duration of 140-150 days.

MATAR AGETA-6

This early maturity group variety has been developed through hybridization (Massey Gem × Harabona) followed by pedigree selection at PAU, Ludhiana and notified by CVRC Notification no. 115 (E) dated 10.02.1996. Plants are dwarf (40 cm), erect, vigorous, with green foliage, bear 12-15 pods (one to two pods per bunch), pods contain upto 6 grain and 44.67% shelling. First picking starts about 50 days after seed sowing with 50% of the total yield. Seeds are rich in protein and dry matter. Plants are tolerant to high temperature. Seeds are green, smooth with slight wrinkles. This variety has been recommended for cultivation in J&K, H.P., Uttaranchal, Punjab, U.P., Bihar, Rajasthan, Gujarat, Haryana and Delhi and has yield potential of 60-65 q/ha.

OREGON SUGAR PODDED (MITHI PHALLI)

This variety is an introduction (Oregon Sugar Podded), recommended by PAU, Ludhiana and notified by CVRC Notification no. 115(E) dated 10.02.1996. It is an edible podded variety suitable for main season, pods are light green, devoid of parchment layer. It is suitable for export purposes. As its shelling is not required, whole pod is eaten so its consumable yield is 2 times more than the shelled pea varieties. Plants are tolerant to powdery mildew under field conditions. This variety has been recommended for cultivation throughout India and has yield potential of 100-110 q/ha green pod.

VIVEK-6 (VL-6)

This medium maturity group variety has been developed through hybridization (Pant Uphar and VL Matar-3) followed pedigree selection at VPKAS, Almora and notified by CVRC Notification no. 98(E) dated 08.02.1997. Plants are dwarf and vigorous with dark green foliage. Pods are smooth, straight, medium-sized (6-7 cm), light green and completely filled. Seeds are semi-wrinkled and greenish white. It is tolerant to cold and moisture stress conditions. This variety has been recommended for cultivation in Uttranchal, Himachal Pradesh, J&K, Haryana, Delhi and Rajsthan and has yield potential of 105-110 q/ha in the duration of 140 and 85 days in mid-hills and plains, respectively.

NDVP-8

This medium maturity group variety has been developed at NDUA&T, Faizabad and notified by CVRC Notification no. 843(E) dated 21.09.1998. Plants are 70-75 cm long, pods 9-10 cm long with 6-8 green and sweet grains. Pods are ready to harvest in 70-80 days after sowing. This variety has been recommended for cultivation in Punjab, Uttaranchal, U.P. and Bihar and has yield potential of 130-160 q/ha.

VL-8

This mid maturity group variety has been developed at VPKAS, Almora and notified by CVRC Notification no. 1135(E) dated 15.11.2001. Pods are green, filled with bold seeds and attractive in colour. This variety has been recommended for cultivation in J&K, H.P. and Uttaranchal and has yield potential of 100-120 q/ha.

VIVEK MATAR-9 (VL-9)

This medium maturity group variety has been developed at VPKAS, Almora and notified by CVRC Notification no. 664 (E) dated 10.05.2005. Plants are medium with dark green foliage. Pods are straight, 6-7 cm long and green. This variety has been recommended for the cultivation in J&K, H.P. and Uttaranchal and has yield potential of 100-105 q/ha.

KASHI NANDINI (VRP-5)

This early variety has been developed through hybridization (P-1542 × VT-2-1) followed pedigree selection at IIVR, Varanasi and notified by CVRC Notification no. 597(E) dated 25.04.2006. Plant height is 47-51 cm, erect and 50% plants bear flowers 34 days after seed sowing foliage is dark green with 7-8 pods per plant. Pods are 8-9 cm long, attractive well-filled with 8-9 seeds, shelling percentage 47-48%. Plants are tolerant to leaf miner and pod borer. This variety has been recommended for cultivation in J&K, H.P., Uttaranchal, Punjab, U.P., Bihar, Jharkhand, Karnataka, Tamil Nadu and Kerala and has yield potential of 110-120 q/ha.

NARENDRA SABZI MATAR-6 (NDVP-12)

This variety has been developed through hybridization (KS-123 × Arkel) followed pedigree selection at NDUA&T, Faizabad and notified by CVRC Notification no. 597 (E) dated 25.04.2006. Plants are green, 45-55 cm tall, flowering starts in 30-35 days, early maturity, first green pod picking in 60-70 days after seed sowing. Pods are ~8 cm long filled with 7-8 green sweet seeds. It is recommeneded for cultivation in Punjab, U.P., Bihar, and Jharkhand and has yield potential of 85-95 q/ha.

NARENDRA SABJI MATAR-4 (NDVP-9)

This variety has been developed at NDUA&T, Faizabad and notified by CVRC Notification no. 597(E) dated 25.04.2006. Plants height 70-75 cm, pods 8-9 cm long with 7-8 green, sweet grains. Plants become ready for harvest in 75-80 days after sowing. It has been recommended for cultivation throughout India and has yield potential of 100-110 q/ha.

NARENDRA SABJI MATAR-5 (NDVP-250)

This variety has been developed at NDUA&T, Faizabad and notified by CVRC Notification no. 597(E) dated 25.04.2006. Plants are 70-75 long with less dark green leaves. Normally, two pods/axil appear, pods are 7-8 cm long with 7-8 green grains. Mature seeds are creamy and wrinkled. Pods are ready to harvest in 80-85 days. It is resistant to powdery mildew and has yield potential of 80-110 q/ha.

KASHI UDAI (VRP-6)

This variety has been developed through hybridization (Arkel and FC-1) followed pedigree selection at IIVR, Varanasi and notified by CVRC Notification no. 597(E) dated 25.04.2006. Plant height 58-62 cm, short internodes with 8-10 pods per plant with dark green foliage and ~50% plants bear flower after 35-37 days of seed sowing. Pods are 9-10 cm long, attractive filled with 8-9 bold seed, shelling percentage 48. This variety has been recommended for the cultivation in Uttar Pradesh and has yield potential of 100-110 q/ha.

KASHI MUKTI (VRP-22)

This early maturity group variety has been developed through hybridization (No. 7 × PM-5) followed pedigree selection at IIVR, Varanasi and notified by CVRC Notification no. 597(E) dated 25.04.2006. Plant height is 50-53 cm and 50% plants bear flowers at 35-36 days after seed sowing (early group). Pods are 8.5-9 cm long, attractive filled with 8-9 bold, soft textured seeds and shelling percentage 48-49. Plants are resistant to powdery mildew. It has been recommended for cultivation in Uttar Pradesh and has yield potential of 120-140 q/ha.

KASHI SHAKTI (VRP-7)

This medium maturity group variety has been developed through hybridization (Hara Bona × NDVP-8) followed pedigree selection at IIVR, Varanasi and notified by CVRC Notification no. 597(E) dated 25.04.2006. Plant height is 90-98 cm and 50% plants bear flowers at

54-56 days after sowing. Plants have dark green foliage with 11-12 pods per plant. Pods are 10-10.5 cm long, attractive filled with 7.5-8.5 bold seed and shelling percentage 48-49. It has been recommended for cultivation in Uttar Pradesh and has yield potential of 140-160 q/ha.

AZAD P-5 (KS-225)

This late maturity group variety has been developed at CSAUA&T, Kanpur and notified by CVRC Notification no. 2035(E) dated 28.11.2006. Plant growth is medium with straight pods full of grains and having extended bearing upto March. It is resistant to powdery mildew and has yield potential of 95-105 q/ha.

SWARNA MUKTI (CHP-2)

This variety has been developed through hybridization followed by pedigree selection at HARP, Ranchi and notified by CVRC Notification no. 2035(E) dated 28.11.2006. Pods are mildly concave, obtuse apexed, long, light green. Shelled seeds are light green and sweet with very good cooking quality. Plants are tolerant to powdery mildew and suitable for September-October sowing. It has been recommended for cultivation in Jharkhand, Bihar and Rajasthan and has yield potential of 120-140 q/ha.

ARKA AJIT (FC-1)

This mid maturity group variety has been developed through back cross and pedigree selection involving the Parents Bonneville, IIHR-209, Freezer-656 (FC-1) at IIHR, Bangalore and notified by CVRC Notification no. 2035(E) dated 28.11.2006. Pods 8-9 cm long, seeds bold, green and sweet, shelling percent 55. Plants are resistant to powdery mildew and rust diseases. It has been recommended for the cultivation in U.P., Rajasthan and Karnataka and has yield potential of 95-100 q/ha in 90 days.

JAWAHAR MATAR-1 (JM-1; GC-141)

This mid maturity group variety has been developed through hybridization (T-19 × Greater Progress) followed by pedigree selection at Research Station, Gwalior and identified for release through AICRP-

VC in 1975. Plants are dwarf (60 cm). Pods are big, attractive, green, 8-9 cm long and contain 8-10 green sweet ovules, with a notch at the base. This variety has been recommended for cultivation throughout the country and has yield potential of 120-125 q/ha of green pod yield.

JAWAHAR MATAR-4 (JM-4; GL-195)

This mid maturity group variety has been developed through hybridization (T-19 × Little Marvel) followed by pedigree selection at Research Station, Gwalior and identified for release through AICRP-VC in 1975. Plants are medium-tall (50-60 cm) and pods are medium (7 cm long), green with 5-6 ovules/pod. Mature seeds are green and wrinkle with 18 g of 100 seed weight. This variety has been recommended for cultivation throughout India and has yield potential of 90-110 q/ha.

PANT UPHAR (IP-3)

This variety has been developed through selection from a germplasm at GBPUA&T, Pantnagur and identified for release through AICRP-VC in 1985. Plants are medium-tall (70 cm) with light green foliage, smaller leaflets, white flower, round pods and wrinkled yellowish seeds with intermediate maturity and first picking at 70-80 days after sowing with 52 % shelling. Plants are susceptible to powdery mildew and resistance to pea stem fly. It has been recommended for cultivation in Punjab, U.P. and Bihar and has yield potential of 100-105 q/ha.

PUNJAB-88 (P-88)

Punjab-88 is an early maturity group variety, developed through hybridization (Pusa-2 × Morrasis-55) followed by pedigree selection at PAU, Ludhiana and identified for release through AICRP-VC in 1985. Plants are dwarf, vigorous, erect with dark green foliage. Pods are dark green, long (8-10 cm) and slightly curved at the center. First picking starts after 100 days of sowing with 47% of shelling. It has been recommended for cultivation in Punjab, U.P., Uttaranchal and Bihar and has yield (green pod) potential of 150-160 q/ha.

PM-2

This variety has been developed through hybridization (Early Badger × IP-3) followed by pedigree selection at GBPUA&T, Pantnagar. Plants are 50-60 cm tall with green pods with 6 ovules. This variety has been recommended for cultivation in J&K, H.P. and Uttaranchal and has yield potential of 50-55 q/ha.

LINCOLN

This is an early maturity group variety, derived from an introduced line (EC-9640) from France at IARI regional station, Katrain and identified for release through AICRP-VC in 1987. Plants are medium-tall and double podded. Pods are dark green, curved, 8-9 cm long, 8-9 ovules/pod with wrinkled mature seeds and good transport quality. In hilly areas, it is prolific bearing. It has been recommended for cultivation in J&K, H.P. and Uttaranchal and has yield green pod potential of 80-100 q/ha.

HISAR HARIT (PH-1)

This variety has been developed at HAU, Hisar and identified for release through AICRP in 1993 and 1995. Plants are semi-dwarf and early with short internodes. It takes ~60 days for first picking. Pods are single to double, large, green, and well-filled seeds and dimpled after drying. This variety has been recommended for cultivation in West Bengal and Assam, Chhatisgarh, Orissa and A.P. and has yield potential of 90-110 q/ha.

NDVP-10

This mid maturity group variety has been developed at NDUA&T, Faizabad and identified for release through AICRP in 1998. Plants are 70-75 cm long, pods 8-9 cm long with 7-8 green, sweet grains. Pods are ready to harvest in 75-80 days after sowing. This variety has been recommended for cultivation in Punjab, Uttaranchal, U.P. and Bihar and has yield potential of 115-125 q/ha.

VRP-2

This is an early maturity group variety developed at IIVR, Varanasi and identified for release through AICRP in 2001. Plants grow upto 50-55 cm height with vigorous dark green foliage. Pods are straight, medium-sized (7-8 cm), light green filled with bold ovules. Flowering starts at 38 days after sowing and green pods may be harvested 55-58 days after sowing. It has been recommended for cultivation in Punjab, U.P. and Bihar and has yield (green pod) potential of 95-105 q/ha.

VRP-3

This mid maturity group variety has been developed at IIVR, Varansi and identified for release through AICRP-VC in 2001. This is the first variety, which floweres between early and mid season, *i.e.*, 45-50 days after sowing. This variety has been recommended for cultivation in J&K, H.P. and Uttaranchal and has yield potential of 80-90 q/ha.

AZAD P-2 (PRS-4)

Azad P-2 has been developed through hybridization (Bonneville × 6587) followed by pedigree selection at CSAUA&T, Kanpur and identified for release through AICRP-VC in 1990. Plants are tall (130-135 cm) and erect having branches with light green foliage and white flowers bearing 20-25 pods (double podded). Pods are medium with 6-7 ovules, nearly straight, light green, smooth and firm. Horticultural maturity duration is of 90-95 days and physiological maturity duration is of 125-130 days. Seeds are wrinkled and brown. It is suitable for cultivation in late sown condition and resistant to powdery mildew. It has been recommended for cultivation in Punjab, U.P., Bihar, Rajasthan, Gujarat, Haryana, Delhi, M.P. and Maharashtra and has yield potential of 115-120 q/ha.

JAWAHAR PEAS-4 (JP-4; JM-6)

This variety has been developed from three way cross [Local Yellow Batri × (6588 × 46-C)] at JNKV, Jabalpur and identified for release through AICRP-VC in 1990. Pods are medium with 5-6 big and green ovules. Seeds are yellow with 15 g of 100 seed weight. Plants are resistant to powdery mildew. This variety has been recommended for cultivation in Punjab, U.P., Bihar, Karnataka, Tamil Nadu and Kerala and has yield potential of 110-120 q/ha.

JP-83

This is a mid maturity group and powdery mildew resistant variety, developed at JNKV, Jabalpur, derived from double cross [(Arkel × JP-829) × (46 C × JP-829)] and identified for release through AICRP-VC in 1992. Plants are dwarf (50 cm). Pods are big, curved with 8 green sweet ovules. This variety has been recommended for cultivation in M.P. and Maharashtra and has yield potential of 120-130 q/ha.

PRIYA (DPP-68)

This variety has been developed at HPKV, Palampur and identified for release through AICRP-VC in 2001. Plants are medium tall, having green wrinkled seeds and light green straight pods containing 7-8 seeds per pod. It takes 100-110 days for first picking. Plants are resistant to powdery mildew disease. It has been recommended for cultivation in Himanchal Pradesh and has yield potential of 150-175 q/ha.

DPP-9411

This mid maturity group variety has been developed through recombination breeding at HPKV, Palampur and identified for release through AICRP-VC in 2002. Plants have deep green dense foliage and are medium tall with an average plant height of 67.6 cm. Pods are deep green, straight, well-filled, medium-sized with pod length of 6-7 cm and containing 7-8 bold grains per pod, sweet with TSS 16.5°B. It has marketable maturity in 130 days in the hills and it is resistant to powdery mildew. It is recommended for cultivation in J&K, H.P. and Uttaranchal and has yield potential of 100-105 q/ha.

COWPEA

Vigna unguiculata
Family: Leguminoceae
Chromosome no.: 2n=22

Vegetable cowpea is popular as leguminous vegetable throughout the world. Cowpea is grown both for its tender pods and also for its dry seeds. Cowpea is more common in northern plains as compared to south. Pods are rich in proteins, vitamin B and minerals. The crop spread from Africa to Asia and Europe through Egypt. It was introduced by the Spaniards into the West Indies in the 16th century and was taken to the United States in around 1700. Africa is considered to be the primary centre of origin. The most probable progenitor of cowpea is var. mensensii. Now cowpea is widely distributed throughout the tropics and subtropics. Cowpea varieties are classified on the basis of duration as early (22-37 days to flower), medium (38-44 days to flower) and late (>45 days to flower). On the basis of growth habit, they are also grouped as pole type (climbing habit) or bush type.

Cowpea plants are vigorously growing annual with deep tap root system. Flowers are white, born as axillary, 2-3 flowers on each peduncle. Flowers open early in morning, close before noon and shed on the same day. Cowpea flowers bloom between 6.30-7.00 h. Anthers mature and start dehiscing about 7-8 hr. before the time of opening of corolla. Pollens remain viable for 42 h at room temperature ($28.5^{o}C$) and relative humidity 91%. When store in a refrigerator at $9^{o}C$ and 0% relative humidity, the period of pollen viability increases to 66 h. Stigma becomes receptive from 12 h before blooming to 6 hr after blooming. Pods are generally 20-30 cm long, cylindrical and slightly curved.

Twenty to twenty-five kg seeds (bushy cultivars) and 10-12 kg seeds (climbing cultivars) is required for planting one hectare area. The summer crop can be sown between mid February to the end of March, whereas, rainy season crop can be sown between mid June to the end of July. Sowing should the done on the ridges in both seasons. Seeds may be dibbled at 10-15 cm distance on the ridges at 50-60 cm apart for bushy varieties. For climbing type cultivars, line-to-line distance of 80-90 cm and seed-to-seed distance of 20-25 cm may be kept; support of bamboos and G.I. wires are provided for better growth and high yield of good quality green pods. It also helps in taking plant protection measures and harvesting. Main objective of breeding cowpea are high yield, photo-insenstiveness, determinate growth habit, and resistance to pests and diseases. A good number of local varieties of vegetable cowpea are available in India. Varieties have also been introduced from South East Asian and Western countries. Selection, mutation and backcross method of breeding have been employed to develop varieties with desirable traits. Golden mosaic virus and pod borer are challenging biotic problems in cowpea.

Varieties

In India, ICAR Institutes and SAUs are working for the improvement of cowpea. The genetic improvement of cowpea is under progress at IIVR (Varanasi), IIHR (Bangalore), HARP (Ranchi), IARI (New Delhi), NDUA&T (Faizabad) and KAU (Vellanikkara). The main objectives of cowpea breeding are non-fibrous pods and resistance against biotic and abiotic stresses. Several varieties of cowpea have been developed as detailed below (for photographs see plate 6).

KASHI SHYAMAL (IVRCP-1)

This variety has been developed at IIVR, Varanasi and notified by CVRC Notification no. 597(E) dated 25.4.06. Plants are dwarf and bushy type (70-75 cm height), suitable for both *Kharif* and *Jaid* seasons, having 3-4 branches, early in flowering (40 days after sowing) with first harvesting after 48 days of sowing producing 35-40 green pods per plant with pod length of 25-30 cm. It is tolerant to Golden mosaic virus. It has been recommended for cultivation in Punjab, U.P., Bihar and Jharkhand and has yield potential of 80-90 q/ha.

KASHI GAURI (VRCP-2)

This variety has been developed at IIVR, Varanasi and notified by CVRC Notification no. 597(E) dated 25.4.06. Plants are bush type, dwarf, photo-insensitive and early variety suitable for sowing in both spring summer and rainy season. It flowers in 35-38 days and pods get ready for first harvest in 45-48 days. Pods are 25-30 cm long, light green as compared to Kashi Shyamal, soft, fleshy and free from parchment layer. Plants are resistant to golden mosaic virus and *Psedocercospora cruenta*. It has been recommended for cultivation in Karnataka, Tamil Nadu and Kerala and has yield potential of 110-130 q/ha.

ARKA SAMRUDHI (IIHR-16)

This early maturing variety has been developed through hybridization (Arka Garima x Pusa Komal) followed by pedigree selection at IIHR, Bangalore and notified by CVRC Notification no. 2035(E) dated 28.11.2006. Plants are erect, bushy, 70-75 cm tall and photo-insensitive. Pods are green, medium thick and long (15-18 cm), tender, fleshy without parchment, good cooking quality. It has been recommended for cultivation in Rajasthan, Gujarat, Haryana, Delhi, M.P. and Maharashtra and has yield potential of 180-190 q/ha in 70-75 days crop duration.

ARKA SUMAN

This variety has been developed developed through hybridization (Pusa Komal x Arka Garima) followed by pedigree selection at IIHR, Bangalore and notified by CVRC Notification no. 2035(E) dated 28.11.2006. Plants are erect, bushy, photo-insensitive with pods above the canopy. Pods are medium long, tender, fleshy, crisp, without parchment good cooking quality. It has been recommended for cultivation in Karnataka. This variety has yield potential of 170-190 q/ha.

ANASWARA

This variety has been developed at KAU, Vellanikkara and notified by CVRC Notification no. 597(E) dated 25.4.06. Vine is semi-determinate type, bears purple flowers and suitable for cultivation as Thadapayar. Pods medium long (28.13 cm) light green, have bold creamy seeds, weight 12.5 g and 19 seeds per pod. This variety has been recommended for cultivation in Kerala and has yield potential of 100-125 q/ha.

SWARNA SUPHALA (CHCP-2)

This variety has been developed through introduction at HARP, Ranchi and notified by CVRC Notification no. 2035(E) dated 28.11.2006. Pods are straight light green, 30-35 cm long with bulged appearance of seeds. Mature dried seeds are bicoloured (mottling of brown and sandalwood). It is suitable for cultivation in summer and rainy seasons. First harvest can be taken 50-55 days after planting. It is resistant to rust and cowpea golden mosaic viruse and tolerant to pod borer during summer under field conditions. It has been recommended for cultivation in Jharkhand, Bihar, Karnataka and Kerala and has yield potential of 210-250 q/ha.

PUSA KOMAL (SEL-1552)

Pusa Komal has been developed at IARI, New Delhi and identified for release through AICRP-VC in 1983. Plants are bushy and flower in 45 days after seed sowing. Pods are light green, tender, 25-30 cm long and suitable for both spring summer and rainy season cultivation. Plants are resistant to bacterial blight. It has been recommended for the cultivation in Uttar Pradesh, Bihar, Punjab and Haryana and has yield potential of 100-115 q/ha.

ARKA GARIMA (SEL-61-B)

This popular variety has been developed through backcross pedigree selection from the cross TUV-762 × *Vigna unguiculata* sub. *sesquipedelis* at IIHR, Banglore and identified for release through AICRP-VC in 1992. Plants are tall, vigorous, bushy, spreading with small vines with light green leaves and purple flowers. Pods are light green, long thick, round, fleshy and stringless; tolerant to heat and drought. It has been recommended for cultivation in M.P., Maharashtra, Karnataka, Tamil Nadu and Kerala and has yield potential of 170-190 q/ha in 90 days of crop duration.

SEL-263

This variety has been developed through pedigree selection at PAU, Ludhiana and identified through AICRP-VC in 1992. Plants are dwarf and can be grown in both spring and rainy season. Pods are green, thick, fleshy, tender and medium (20 cm long). Plants are resistant to mosaic and golden mosaic viruses. This variety has been recommended for cultivation in states of Punjab, U.P. and Bihar and has yield potential of 200-220 q/ha.

SEL-2-1

This variety has been developed at NDUA&T, Faizabad and identified for release through AICRP-VC in 1993. Plants are 70-75 cm tall, pods are green, length 25-30 cm, seed black; susceptible to *Pseudocercospora* and viruses. This variety has been recommended for cultivation in Punjab, U.P. and Bihar and has yield potential of 125-150 q/ha.

NARENDRA LOBIA-2 (NDCP-13)

This variety has been developed through hybridization (Sel-2-1 × Red Seeded) followed by pedigree selection at NDUA&T, Faizabad and identified through AICRP-VC in 2002. Plants are bushy, early flowering, dark green and 28 cm long. First picking starts after 50 days of seed sowing and suitable for kharif and zaid cultivation. It has been recommended for cultivation in West Bengal, Assam, North East States, Andmand & Nicobar Islands, Punjab, U.P., Bihar, Jharkhand, Karnataka, Tamil Nadu and Kerala and has yield potential of 75-100 q/ha.

KASHI KANCHAN (IIVRCP-4)

This variaty has been developed at IIVR, Varanasi and identified for release through AICRP-VC in 2007. It is a bush type and dwarf (50-60 cm height), photo-insensitive and early variety suitable for sowing in both spring-summer and rainy seasons. It flowers in 40-45 days and pods get ready for harvest in 50-55 days. It produces 40-450 pods per plant of 30-35 cm long. The pods are dark green, tender, pulpy with less fibre and free from parchment layer. The varity is resistant to golden mosaic virus and *Pseudocercospora cruenta*. It has been recommended for cultivation in Uttar Pradesh, Chhattisgarh, Madhya Pradesh and Maharashtra and has green pod yield potential of 150-200 q/ha.

VR-5

This variaty has been developed at IIVR Varanasi and identified for release through AICRP-VC in 2008. It is a bush type and early maturing, suitable for spring summer and rainy seasons sowing. Flowering starts 35-40 days after sowing and produces 40-45 pods per plant. It has been recommended for cultivation in Uttar Pradesh, Chhattisgarh, Madhya Pradesh and Maharashtra and has green pod yield potential of 150-175 q/ha.

FRENCH BEAN

Phaseolus vulgaris
Family: Leguminoceae
Chromosome no.: 2n=22

French bean (syn. kidney bean, haricot bean, snap bean, navy bean, Farash bean) is an important vegetable used for its green tender pods. Southern Mexico and Central America are considered to be the primary center of origin of French bean, while Peruvian-Ecuadorian-Bolivian areas are considered to be secondary centre of origin. The progenitor of French bean is *Phaseolus aboriginens*. *P. vulgaris*, *P. coccineus*, *P. lunatus* and *P. acutifolius* var. *latifolius* are cultivated species of genus *Phaseolus*. All these species are self-pollinated, whereas *P. coccineus* is cross-pollinated. French bean cultivars are generally grouped on the basis of their growth habit. There are 3 type of French bean, *viz*., bush type (short internode), semi-pole type (longer internodes than those in bus type), and pole type (longer internodes than that of semi-pole type). Most of the French bean varieties are day neutrals except few short day types. Day temperature of 15-25°C is optimum to harvest the maximum yield. The suitable soil pH is 5.3-6.0. Present improved varieties of cowpea can be grown round the year. In hilly regions, sowing can be done from March to first week of May. The seed rate of bush type varieties (65 kg/ha) are more than pole type (20-30 kg/ha). Pole type varieties need to be staked on bamboo or iron poles. The first picking generally starts after 45 days of sowing. Pole type varieties are generally long duration crops. The main breeding methods are introduction, mutation and back crossing.

Varieties

The main objectives of French bean breeding are attractive pods without parchment payer, bushy growth habit, resistant to biotic stress coupled with high yield. IIHR (Bangalore), IIVR (Varanasi), IARI Regional Station (Katrain), GBPUA&T (Pantnagar) are main research centres where genetic improvement is under progress. Several varieties have been developed and described in detail as under (for photographs see plate 7).

KASHI PARAM (IVFB-1)

This variety has been developed through pureline selection at IIVR, Varanasi and notified by CVRC Notification no. 2035(E) dated 28.11.2006. Plants are determinate (70 cm long) with dark green leaves. Pods are fleshy, length 14.7 cm, round, dark green, seed colour dark brown to green. This variety has been recommended for cultivation in J&K, H.P. and Uttaranchal and has yield potential of 120-140 q/ha.

ARKA SUVIDHA (IIHR-909)

This variety has been developed through hybridization (Blue Crop × Contender) followed by pedigree selection at IIHR, Bangalore and notified by CVRC Notification no. 2035(E) dated 28.11.2006. Plants are bushy and photo-insensitive. Pods are straight, oval, light green, fleshy and stringless. It has been recommended for cultivation in J&K, H.P. and Uttaranchal and has yield potential of 180-195 q/ha in 70 days of crop duration.

AZAD RAJMAH-1

This variety has been developed at CSAUA&T, Kanpur and notified by CVRC Notification no. 597 (E) dated 25.4.2006. Plants are dwarf with dark green foliage. Pods are light green, smooth, tender stringless and length 12-15 cm. It has yield potential of 75-80 q/ha.

PHULE SUYASH (GK-7)

This variety has been developed through selection from Shahpur Local cultivar at MPKV, Rahurui and notified by CVRC Notification no. 597(E) dated 25.04.2006. Pods are attractive tender, straight to slightly curved, smooth and green. Plants are highly tolerant to wilt- and bean mosaic virus. It has been recommended for the cultivation in Maharashtra and has yield potential of 230-245 q/ha in 70-80 days of crop duration.

VL BAUNI-1

This variety has been developed at VPKAS, Almora and identified for release through AICRP-VC in 1985. Plants are dwarf (30-40 cm) having white flowers with purple tinge. Pods are medium long, light green, attractive, stringless and fleshy. First picking starts at 45-60 days. Plants are moderately susceptible to foliar diseases. This variety has been recommended for cultivation in Uttaranchal and has yield potential of 110-111 q/ha.

ARKA KOMAL (SEL-9)

Arka Komal has been developed through selection from an Australian Introduction (IIHR-60) at IIHR, Bangalore. Plants are erect, bushy with bright green stem and foliage. Pods are straight, flat, tender, green with oblong, large and light brown seeds with good keeping and cooking qualities. This variety has been recommended for cultivation in J&K, H.P., Uttaranchal, M.P., Maharashtra, Karnataka, Tamil Nadu and Kerala and has yield potential of 180-195 q/ha pod yield and 25-30 q/ha seed yield.

PANT ANUPMA (UPF-191)

Plant Anupma has been developed through selection from line UPF-191 at GBPUA&T, Pantnagar. Plants are bushy, dwarf, upright green foliage having concentrated fruiting at mid height. Pods are fully covered by leaf canopy, thus protected from sun light; first picking starts 55-56 days after seed sowing. It is moderately resistant to bean common mosaic and angular leaf spot. This variety has been recommended for cultivation in Punjab, Uttaranchal, U.P., Bihar, M.P. and Maharashtra and has yield potential of 90-100 q/ha.

SWARNA PRIYA (CH-812)

This variety has been developed through pure line selection at CHES, Ranchi and identified through AICRP-VC in 2001. Pods are flat, green, fleshy having good cooking quality. Dried seeds are bold and maroon in colour, suitable for 'daal' preparation. First harvest can be taken 50-55 days after planting. Suitable for mid September sowing and it is tolerant to powdery and downy mildew and resistant to leaf minor under field conditions. It has been recommended for cultivation in Jharkhand, Sikkim, Meghalaya, Manipur, Nagaland, Mizorum, Tripura, Arunachal Pradesh, Madhya Pradesh, Maharashtra and Andaman and Nicobar Islands and has yield potential of 120-140 q/ha.

SWARNA LATA (CH-819)

This variety has been developed through pure line selection at CHES, Ranchi and identified for release through AICRP-VC in 2001. This is suitable for mid June and September sowing. Pods are medium long, fleshy and stringless with good cooking quality. First harvest can be taken 55-60 days after planting. It is tolerant to powdery and downy mildew and resistant to leaf minor under field conditions. This variety has been recommended for cultivation in Jharkhand, J&K, Himachal Pradesh and Uttaranchal and has yield potential of 120-140 q/ha.

INDIAN BEAN

Dolichos lablab
Family: Leguminoceae
Chromosome no.: 2n=22

Dolichos bean (syn. hyacinth bean, bonavist bean, Indian bean, Egyptian bean, field bean, Australian pea, Indian butter bean, sem and baby salad bean) is an important vegetable of India, grown for their green, and tender pods. It is a cool season and drought tolerant crop of Indian origin. It is commercially cultivated throughout the tropics and temperate region of Asia, Africa and America. Although plants are herbaceous perennial but they are cultivated as annual. The anthesis of flowers takes place 9 to 17 h, whereas anther dehiscence takes between 5-14 h. Pollen fertility and stigma receptivity is only on the day of anthesis.

Dolichos bean is strictly a self-pollinated crop. Plants start fruiting from the beginning of cool season. Now photo-insensitive varieties are available, which can be grown throughout the year. The seed rate of dolichos bean is 50-60 kg/ha, they are usually inoculated with *Rhizobium* sp. in order to facilitate quick nodulation of roots, which fix nitrogen from atmosphere. Climbing type varieties are allowed to grow on pandal system. Lack of moisture during flowering and pod development reduce the yield. Pod borer, aphids are plant hoppers are major insects, whereas powdery mildew, rust and *Phytophthora* are serious diseases of dolichos bean. First picking starts after two months of sowing in bush type and three months in pole type.

Varieties

The most commonly used breeding methods are pure line selection, bulk method, pedigree selection, mass selection, and mutation breeding. The breeding objectives are bush and dwarf plant types with quality pod characters, earliness and resistance to bean mosaic virus. The improvement works on *Dolichos* is under progress at IIHR (Bangalore), IARI (New Delhi), IIVR (Varanasi), HARP (Ranchi), and NDUA&T (Faizabad). Several varieties of Indian bean have been developed and described in detail as under (for photographs see plate 8).

HEBBAL AWARAI-3

This variety has been developed through hybridization (Hebbal Awarai-1 × US-67-13) followed by pedigree selection at IIHR, Bangalore and notified by CVRC notification no. 19(E) dated 14.01.1982. Plants are 65.75 cm tall, erect, determinate and photo-insensitive. Flowers are white, round and small. It matures 90-110 days after sowing and has yield potential of 180-200 q/ha of green pods.

CO-1

This variety has been developed through selection from a germplasm (*D. lablab* var. *typicus*) at TNAU, Coimbatore and notified by CVRC notification no. 19(E) dated 14.01.1982. It is earlier in maturity than other pole type varieties. The pods are ready for harvest in 80-85 days after sowing. The total crop duration is ~165 days. Pods are dark green, flattish, slightly curved, tender and fleshy with better consumer's preference. Mature seeds are black with white hylem with 100 seeds weight of ~50 g. This variety has yield potential of 180-200 q/ha of green pods.

WAL KONKAN-1

This variety has been developed through hybridization (Wal-2 K-2 × Wal-125-36) followed by pedigree selection at KKV, Dapoli and notified by CVRC Notification no. 295(E) dated 09.04.1985. Plants are 10-45 cm tall, bushy with long tendrils. Leaves are dark green and smooth. Flowering starts 55-60 days after sowing. Seeds are bold with brown coat colour. Seeds contain 14.68% protein. Plants are resistant to yellow mosaic virus. It matures 110-115 days after sowing and has yield potential of 70-90 q/ha.

CO-2

This variety has been developed through hybridization (CO-8 × CO-1) followed by a pedigree selection at TNAU, Coimbatore and notified by CVRC Notification no. 540(E) dated 24.07.1985. Plants are 60 cm tall, photo-insensitive, erect, and bushy with 5-6 branches. Flowers are pinkish purple at blossom, bluish purple on fading and borne on axillary racemes. Pods are flat, green and four seeded. Seeds are medium, ovate and black. This variety has yield (green pods) potential of 180-200 q/ha.

KONKAN BHUSHAN

This variety has been developed through selection from a germplasm at KKV, Dapoli and notified by CVRC notification no. 617(E) dated 17.08.1993. It is a bush type variety of 60-75 cm height. Pods are tender, green and stringless. Single plant produces 125-180 pods each of 3-4 g. The pod length is 10 cm long and diameter is 1.5 cm. It has proved to be adaptable in varying soil and climates with fairly economic yield. It matures in 55-60 days after sowing and has yield potential of 80-100 q/ha.

PHULE GOURI (RHRWL-1)

This variety has been developed through selection from a germplasm at MPKV, Rahuri and notified by CVRC Notification no. 2035(E) dated 28.11.2006. Pods are whitish green, tender, flat and slightly curved, and pod length 7-9 cm. Fruits are smooth and attractive green with white strips at apical end and bigger in size. Plants are tolerant to pod borer, anthracnose, leaf spot and leaf minor under field condition. This variety has yield potential of 330-345 q/ha in 180 days of crop duration.

SWARNA UTKRIST

This variety has been developed through pure line selection at HARP, Ranchi and notified by CVRC Notification no. 2035(E) dated 28.11.2006. Pods are straight, flat, green & fleshy pods (10-12 cm) having very good cooking quality. Mature seeds are light brown. Plants are tolerant to anthracnose, mosaic viruses and pod borer under field condition. First harvest can be taken 110-120 days after planting. It has been recommended for December-January sowing in U.P., Jharkhand, Bihar and Punjab and has yield potential of 375-400 q/ha.

Chapter 4

Bulb Vegetables

ONION

Allium cepa
Family: Lilliaceae
Chromosome no.: 2n=16

Onion is an important vegetable grown for its bulb, which is used in cooked vegetables, consumed as salad or pickled. India ranks second in term of area and production in the world after China. Maharashtra, Gujrat, Karnataka, Orissa, Andhra Pradesh, Tamil Nadu, Uttar Pradesh, Rajasthan and Bihar are the major onion growing states in India. Onion posses many medicinal properties and has been used to treat common cold, measles and rheumatism. The primary center of origin of onion is central Asia and near east is considered to be the secondary center of origin. The inflorescence in onion is umbel with 50-2000 flowers.

It is a cool season crop and flowering is usually affected by day length. Therefore, the varieties are classified as *rabi* or *kharif* season. In most of the commercial varieties, relatively high temperature as well as long photoperiod is essential for bulb formation. Sandy loam soil with pH 5.8-6.5, rich in humus with good drainage is highly preferred for onion cultivation. The best time for sowing early *kharif* crop is April-May in South, and May-June for northern states. *Rabi* season onion can be sown in the month of September-October in south, November and December in Maharastra and part of Gujarat, October-November in northern states and NEH region.

Approximately, 10 kg seeds are required to transplant one hectare area. Nursery is generally sown on raised bed. Generally, soil of the nursery is treated with thiram or captan @ 0.2% or 4-5 g/m^2 area. Seeds are also treated with thiram @ 2-3 g/kg seed before sowing. Nursery should be free from weeds during its early growth. For *kharif* onion, the nursery becomes ready for transplanting after 6-7 weeks old and rabi season, the nursery should be 7-8 weeks old. If overaged seedlings are transplanted, premature bolting takes place, which is undesirable. For direct sowing of small-sized bulb varieties, *i.e.*, Agrifound Red, 22.5 kg seeds/ha is required.

Seedlings are transplanted in flat beds. *Kharif* onions are transplanted on both sides of ridges in highly rainfall-prone areas. The spacing is kept 15 x 10 cm from row-to-row and plant-to-plant. The ridges should be prepared at a distance of 30-45 cm, big size onion varieties are grown through sets. The size of sets should be 1.5-2 cm for getting good quality bulb. Frequent and light irrigations at weekly intervals, results good bulb development and increased yield. Bulb formation is critical stage for irrigation. When top mature and start falling in rabi season, irrigation should be stopped, while in kharif it should be stopped 10 days before harvesting. In northern India, onion is usually grown in winter season (rabi). Now, kharif onion in northern India is becoming very popular due to high remunerative price. In northern plain, for successful cultivation of crop, seed is sown in May/June and transplanted in July-August. Bulb becomes ready for harvesting in December-January. The most common kharif varieties are N-53, Agrifound Dark Red, Baswant-780, and Arka Kalyan.

Varieties

At present, ICAR Institutes, NHRDF (Nasik) and NRCO&G (Pune) are involved in the development of onion varieties and hybrids. Introduction, mass selection, selfing and massing, and hybridization are mainly used for development of onion varieties. Several varieties suitable for different seasons have been developed and described in detail as under (for photographs see plate 9).

CO- 2

This variety is a selection from a germplasm (CS-911) developed at TNAU, Coimbatore and notified by CVRC Notification no. 13 dated 19.12.1978. Plant height is ~30 cm. Leaves are green and cylindrical. The number of shoots per plant ranges from 7 to 9. Bulbs are crimson in colour and moderately bigger in size. Mean bulb weight per plant ranges from 60 to 65 g with an average of 9 bulbs per plant. The weight loss during storage is lesser than variety CO-1. This variety is pungent with high total soluble solids content (12° brix) and shorter in duration than CO-1. It has yield potential of 120-140 q/ha in a crop duration of 65 days.

PUNJAB SELECTION

This variety has been developed through mass selection from an indigenous collection at PAU, Ludhiana and notified by CVRC Notification no. 361(E) dated 30.06.1973. Plants are 55-65 cm tall with 6-9 leaves per plant. Bulbs are globular, red, 5-6 cm in diameter, weight of 50-70 g, firm and have good keeping quality. It has been recommended for cultivation in Punjab, U.P., Bihar, M.P., Maharashtra, Karnataka, Tamil Nadu and Kerala and has yield potential of 300-325 q/ha.

HISAR-2

This variety has been developed at HAU, Hisar and notified by CVRC Notification no. 786 dated 02.02.1976. Bulbs are bronze red and flatish globular shape. Bulbs have very good keeping quality. This variety is tolerant to salinity. It has been recommended for cultivation in Haryana and Punjab and has yield potential of 310-330 q/ha in 130 days of crop duration.

PUSA RATNAR

This variety has been developed through single plant selection at IARI, New Delhi and notified by CVRC Notification no. 13 dated 19.12.1978. Plants are 30 cm tall and dark green leaves with waxy bloom. Bulbs are bronze deep red, more exposed (above the ground at maturity), obviate to flat-globular, less pungent with 12% TSS and drooping neck with good storage quality. It is recommended for cultivation throughout the country and has yield potential of 325-350 q/ha in 150 days of crop duration.

PUSA RED

This variety has been developed from an indigenous collection at IARI, New Delhi and notified by CVRC Notification no. 13 dated 19.12.1978. Plants are 55-65 cm tall with 6-9 leaves per plant. Bulbs are bronze red in colour, flat globular in shape with 12-13% TSS content, weight 70-90 g, measure 5-6 cm across, less pungent and quite firm with good keeping quality. It has been recommended for cultivation in Punjab, U.P. and Bihar, Rajasthan, Gujarat, Haryana, Delhi, Karnataka, Tamil Nadu, Kerala and has yield potential of 250-300 q/ha in 125-140 days of crop duration.

CO- 3

This variety is a clonal selection (CS-450) developed at TNAU, Coimbatore and notified by CVRC Notification no. 19 (E) dated 14.01.1982. Leaves length 40 cm, light green, erect and cylindrical. The number of shoots ranges from 6 to 8. The bulbs are pink in colour and bigger in size. The bulb length and girth are 3.5 cm and 8.3 cm, respectively. The mean bulb weight is 75 g. Number of bulbs per plant ranged from 8 to 10. Bulbs are bigger than CO-2 with good consumer appeal and contain 17.5% dry matter, 0.53% sulphur and 13.0% TSS. Bulbs can be stored well for 120 days devoid of sprouting in well-ventilated store room. It is moderately resistant to thrips and has yield potential of 160-180 q/ha in crop duration of 65 days

KALYANPUR RED ROUND

This variety has been developed at CSAUA&T, Kanpur and notified by CVRC Notification no. 2(E) dated 03.01.1983. Bulbs are bronze red, globular, pungent with and 13-14% TSS. This variety has yield potential of 250-300 q/ha in 150 days of crop duration.

CO-4

This variety has been developed through hybridization (AC-863 × CO-3) followed by pedigree selection at TNAU, Coimbatore and notified by CVRC Notification no. 596(E) dated 13.08.1984. Leaves are light green, erect and cylindrical. The number of leaves per plant varies from 8 to 13 with an average weight of 90 g per clump. Bulbs are attractive light pink, bold in size with good consumer appeal. The bulbs stored upto 150 days. Occasional incidence of *Alternaria* blight has been observed in this variety. It is moderately resistant to thrips and has yield potential of 200-225 q/ha in a crop duration of 65 days.

BASWANT-780

This variety has been developed at IIHR, Bangalore and notified by CVRC Notification no. 915(E) dated 06.11.89. It is grown in *kharif* season in Maharashtra. This is a big size onion variety grown in *rabi* season. Bulbs are attractive red in colour, globular in shape and mildly pungent, less bolting with twins. It becomes ready for harvest in 100-110 days and has yield potential of 200-250 q/ha and suitable for export purpose.

PUSA MADHVI (LINE-120)

This variety has been developed at IARI, New Delhi and notified by CVRC Notification no. 915(E) dated 06.11.1989. Plant height is 55 cm. Bulbs are light red, flattish round with TSS 11-13%. Bulbs mature in 130-145 days. It is recommeneded for the cultivation in Madhya Pradesh and Maharashtra and has yield potential of 300-325 q/ha in 135 days of crop duration.

AGRIFOUND DARK RED

This variety has been developed through mass selection from a local collection of *kharif* onion at NHRDF, Nasik and notified by CVRC Notification no. 1135(E) dated 01.12.1998. Bulbs are dark red, globular, 4-6 cm in size, moderately pungent with 12-13% of TSS and thin skin with average keeping quality. This variety has been recommended for cultivation during *kharif* season in Punjab, U.P. and Bihar and has yield potential of 375-400 q/ha in 95-100 days of crop duration.

ARKA NIKETAN (SEL-13)

This variety has been developed through mass selection from a local collection (IIHR-153) at IIHR, Bangalore and notified by CVRC

Notification no. 386(E) dated 15.05.1999. Plants are of 60-70 cm height with 8-10 leaves per plant. Bulbs are globular, pink, outer scales arranged in tight concentric ring weight of 100-180 g, highly pungent with 12-14% TSS and thin neck. It has good storage quality and can be stored up to 5 months at ambient temperature. It is suitable for both rabi and kharif season cultivation. It has been recommended for cultivation in M.P. and Maharashtra and has yield potential of 420-450 q/ha in 145 days of crop duration with seed yield of 80 q/ha.

ARKA KALYAN (SEL-14)

This variety has been developed through mass selection from a local collection (IIHR-145) at IIHR, Bangalore and notified by CVRC Notification no. 386(E) dated 15.05.1999. Plants are of 70-90 cm height, Bulbs are globular, pink, flesh succulent, concentric internal scale, weight of 130-190 g with 11-13% TSS and moderately resistant to purple blotch. It is suitable only for kharif season cultivation. It has been recommended for cultivation in Punjab, U.P., Bihar, Rajasthan, Gujarat, Haryana, Delhi, M.P., Maharashtra, Karnataka, Tamil Nadu and Kerala and has yield potential of 470-500 q/ha in 140 days of crop duration and seed yield of 80 q/ha.

AGRIFOUND LIGHT RED

This variety has been developed at NHRDF, Nasik and notified by CVRC Notification no. 115(E) dated 10.02.1996. It has been derived through mass selection from a local stock of *rabi* season cultivar, grown in Dindori area of Nasik. Bulbs are light red, globular, tight skin, 4-6 cm in size with 13% of TSS having good keeping quality. It has been ecommended for cultivation during *rabi* season in M.P. and Maharashtra and has yield potential of 300-325 q/ha in 160-165 days of crop duration.

VL PIAZ-3

This variety has been developed at VPKAS, Almora and notified by CVRC Notification no. 793(E) dated 22.11.1991. Plants are 50-70 cm tall with deep green waxy foliage. Bulbs are red, medium, globular and pungent. This variety has been recommended for cultivation in Uttaranchal and has yield potential of 250-275 q/ha in 145 days of crop duration.

PUNJAB NAROYA (PBR-5)

This variety has been developed at PAU, Ludhiana and notified by CVRC Notification no. 115(E) dated 10.02.1996. PBR-5 has been derived through selection from a germplasm collected from Maharashtra. Shoots are medium-tall and green. Bulbs are red, medium to large, round with thin neck. Plants are tolerant to purple blotch, thrips and *Heliothis*. It has been recommended for cultivation in Rajasthan, Gujarat, Haryana and Delhi and has yield potential of 375-400 q/ha in 125 days of crop duration.

PUNJAB WHITE

This variety has been developed at PAU, Ludhiana and notified by CVRC Notification no. 843(E) dated 21.09.1998. Bulbs are suitable for dehydration, average TSS 15%, bulbs medium large and round with narrow neck. It has been recommended for cultivation in Punjab and Haryana and has yield potential of 310-330 q/ha.

PHULE SUVARNA (RHR-87015)

This variety has been developed through selection from a cross (RHR-200 × RHR-155) at MPKV, Rahurui and notified by CVRC Notification no. 348(E) dated 08.05.2001. Bulbs are yellow, less pungent, TSS 11-12%, and suitable for export. It is moderatly resistant to purple blotch and has yield potential of 300-310 q/ha in kharif and 290-300 q/ha in rangda season.

ARKA PITAMBER

This variety has been developed at IIHR, Bangalore and notified by CVRC Notification no. 2035(E) dated 28.11.06. It is a short-day yellow onion released for cultivation in Karnataka. Bulb are uniformly yellow in colour with perfect globe shape, very firm, single centered, thin neck, mild pungent and medium-sized with good keeping quality. It is suitable for the export to European countries, USA and Japan as recommended by NHRDF, Nasik and has yield potential of 350-375 q/ha.

ARKA BINDU

This variety has been developed from local rose onion collection IIHR-402 through mass selection at IIHR, Bangalore and notified by

CVRC Notification no. 2035(E) dated 28.11.2006. Its bulbs are deep red, small-sized, flatish-globe and uniform in shape, size and maturity. It has high TSS, pungency and dry matter content. Bulbs are firm with good storage quality. It has been recommended by NHRDF for export and has yield potential of 325-350 q/ha.

AKOLA SAFED

This variety has been developed at PDKV, Akola and notified by CVRC Notification no. 597(E) dated 25.4.2006. Plant height ~80 cm, foliage dark green with 8-10 leaves. Bulbs are white with neck thickness of 1 cm, and diameter of 4.9-6.5 cm. Bulbs attain commercial maturity in 180-185 days after sowing. Average bulb weight is 70-80 g with 11–12 % TSS. This variety has yield potential of 275-300q/ha.

PHULE SAMARTH (S-1)

This variety has been developed through selection from a germplasm (Sangamner) at MPKV, Rahurui and notified by CVRC Notification no. 597(E) dated 25.04.2006. Bulbs are lusterious and dark red. It is moderatly resistant to purple blotch and has yield potential of 300-310 q/ha in kharif and 500-520 q/ha in rangda season.

RAJASTHAN ONION-1 (RO-1)

This variety has been developed at RARS, Durgapura and notified by CVRC Notification no. 642(E) dated 31.05.2004. Plant height is 62.5 cm, bulb weight 65.9 g, diameter 4.6 cm, shape flattened globe, copper red, pungency mild with 10.5% TSS and 6.5% dry matter. It is resistant to purple blotch and has yield potential of 325 -360 q/ha in 130-140 days of crop duration.

APRITA (RO-59)

This variety has been developed at RARS, Durgapura and notified by CVRC Notification no. 664(E) dated 11.05.2005. Plant height is ~45 cm, bulb weight 74.6 g, diameter 5.2 cm, shape flattened globe, red, pungency mild, TSS 10.2% and dry matter content 7.8%. It is resistant to purple blotch and has yield potential of 350-400 q/ha in 122-126 days of crop duration.

PUSA WHITE FLAT (SEL-131)

This variety been developed at IARI, New Delhi and identified for release through AICRP-VC in 1977. Bulbs are medium-large, flattish round with attractive white colour, 12-14% TSS and good keeping quality. This variety has yield potential of 300-325 q/ha 120-130 days of crop duration.

N-257-9-1

This variety has been developed through selection from a local collection at MPKV, Rahuri and identified for release through AICRP-VC in 1985. Bulbs are globular and white, suitable for rabi season cultivation. It has been recommended for cultivation in M.P. and Maharashtra and has yield potential of 250-275 q/ha in 120 days of crop duration

N-2-4-1

This variety has been developed at MPKV, Rahuri and identified for release through AICRP-VC in 1985. Plants are of 40-45 cm height. Bulbs are globular (6.1 cm diameter), brick red with 12-13% TSS, pungent and firm with good keeping quality. It is suitable for rabi season. It has been recommended for cultivation in Punjab, U.P., Bihar, M.P. and Maharashtra and has yield potential of 300-325 q/ha in 140 days of crop duration

PUNJAB RED ROUND

This variety has been developed at PAU, Ludhiana and identified for release through AICRP-VC in 1977. It is an early maturing variety, bulbs are shining red, globular, medium with thin neck. It has been recommended for cultivation in Punjab, U.P. and Bihar and has yield potential of 280-300 q/ha.

HOS-1

This variety has been developed for rabi season at HAU, Hisar and identified by AICRP-VC in 2006. Bulbs are bronze red, globular with 11-13% TSS. It has been recommended for cultivation in U.P., Bihar,

Jharkhand and Uttaranchal and has yield potential of 200-250 q/ha in 165 days of crop duration.

LINE-28

This variety has been developed for rabi season at NHRDF, Nashik and identified by AICRP-VC in 2006 through selection. Its leaves are dark green, erect with a thin waxy coating. Bulbs are dark, scarlet red, globular round, reddish with thick inner scales, fleshy thick red scales and 13-14% TSS. Bulbs have 2-3 outer scales and moderately pungent in taste. It has been recommended for cultivation in northern India and has yield potential of 250-285 q/ha.

BHIMA RAJ (B-780-5-2-2)

This variety has been developed through bulb-to-row selection at NRC on Onion and Garlic, Pune and identified by AICRP-VC in 2007. Bulbs are oval, dark red, striped having thin neck and 10.5% TSS. It is suitable for rabi season. It has been recommended for cultivation in Haryana, Delhi, Rajsthan and Gujrat and has yield potential of 250-300 q/ha.

GARLIC

Allium sativum
Family: Alliaceae
Chromosome no.: 2n=16

Garlic is the second most important bulb crop grown throughout the country. Central Asia, where it extends from mountainous (sourthern), Turkmenia north eastward in Pamir-Alai considered to be the primary center of origin. It is consumed as green as well as dried in the form of spice. It is used in various Indian preparations as flavoring ingredient. Garlic is also used to disguise the smell and flavour of salted meat and fish. Garlic powder is now exported to different countries. Garlic is a rich source of carbohydrate, proteins and phosphorus. Several medicinal properties of garlic have been documented. It is carminative and as a gastric stimulant help in digestion and absorption of food. Important garlic growing states in India are Gujarat, Madhaya Pradesh, Orissa, Rajasthan, Maharastra and Uttar Pradesh. Gujarat has the largest area and production followed by Madhaya Pradesh, Orissa and Maharastra.

Well fertile sandy loam soil having pH 6.0-7.0 is the best for garlic cultivation. It is a frost tolerant crop, requires cool and moist period during vegetative growth and dry period during bulb maturity. Most of the cultivated varieties of garlic in India are short day. Cloves are used as a seed and about 500 kg cloves of 8-10 mm dia are required to plant one hectare area. The distance of sowing varies from region to region. In general, 15x10 cm spacing is recommended to get the quality and appropriate size colves. The sowing of garlic differs from one zone to another. In Madhya Pradesh, Maharastra, Karnataka and Andhra Pradesh, it can be sown in month of August-October, whereas, September-November is the best for Northern India. October-November is the best time for sowing in Gujarat, West Bengal and Orissa states. April-May is the best time of sowing at hills. Bacterial and *Fusarium* bulb rotting are serious diseases, while onion thrips is another major insect. Bulbs are ready for harvesting after 4-5 months of sowing. Mutations arising during long vegetative phase provide a chance to select variants.

Varieties

In the case of garlic, variants have been selected for plant size, yield, solids content, time of maturity, shape and colour of cloves.

At present genetic improvement is being carried out at NRCO&G (Pune), NHRDF (Nashik) and state agricultural universities. Introduction, mass selection, selfing and massing, and hybridization are mainly used for development of garlic varieties and several varieties have been developed and described in detail as under (for photographs see plate 10).

AGRIFOUND WHITE (G-41)

This variety has been developed through mass selection from local collection of Bihar Sharif at NHRDF, Karnal and notified by CVRC Notification no. 280(E) dated 13.04.1989. Bulbs are compact, silvery white skin with creamy flesh, diameter 3.5 to 4.5 cm and length with 12-15% TSS, 42.78% dry matter. Cloves are large, elongated with 20-25 cloves/bulb. It has good storage quality and susceptible to purple blotch and *Stemphylium* blight. It has been recommended for cultivation in Punjab, U.P., Bihar, M.P. and Maharashtra and has yield potential of 130-150 q/ha.

YAMUNA SAFED (G-1)

This variety has been developed through mass selection from local collection at NHRDF, Karnal and notified by CVRC Notification no. 527(E) dated 16.08.1991. Bulbs are compact, silvery white skin, creamy flesh, TSS 38%, dry matter 39.5%, 4.0-4.5 cm diameter, 12-15 cm^2 size index with 25-30 cloves per bulb. Cloves are sickle shaped and 0.8-1.0 cm diameter with good storage quality. It is tolerant to purple blotch, *Stemphylium* blight and thrips. It is recommended for cultivation in Rajasthan, Gujarat, Haryana, Delhi, M.P. and Maharashtra and has yield potential of 150-175 q/ha.

YAMUNA SAFED-2 (G-50)

This variety has been developed through mass selection from a local collection of Karnal at NHRDF, Karnal and notified by CVRC Notification no. 115(E) dated 10.02.1996. G-50 was developed at NHRDF, Nasik. Bulbs are compact, white creamy flesh, 10 bulb weight 160-240 g; T.S.S. 38-40%, dry matter 40-41%, diameter 3.5-4.0 cm, number of cloves 35-40, 0.75-1.0 cm diameter, 1.75-2.5 cm^2 clove size index. It has been recommended for cultivation in Punjab, U.P. and Bihar and has yield potential of 150-200 q/ha.

GHC-1

This variety has been developed at HPKV, Palampur and notified by CVRC Notification no. 115(E) dated 10.02.1996. Bulbs are large with 12-14 cloves. Cloves are pungent and easily separated from bulb and has yield potential of 275-300 q/ha.

YAMUNA SAFED-3 (G-282)

This variety has been developed at NHRDF, Karnal and notified by CVRC Notification no. 1052(E) dated 26.10.1999. Bulbs are creamy white and bigger-sized (5-6 cm dia), size index 27-29 cm^2. Clove diameter is 1.2-1.5 cm with 15-16 cloves/bulb and 38.42% TSS, 39-43% dry matter. This variety is suitable for export and has yield potential of 175-200 q/ha.

YAMUNA SAFED-4 (G-323)

This variety has been developed at NHRDF, Karnal and notified by CVRC Notification no. 597(E) dated 25.04.2006. Plants vigorous with wider green leaves, bulbs silvery white bigger sized (3.5-4 cm), bulb size index 12-15 cm^2, clove diameter 1.2-1.25 cm, clove size index 2.5-3.0 cm^2, 25-30 cloves per bulb having 40-42% TSS. It has been recommended for cultivation in Rajasthan, Gujarat, Haryana and Delhi and has yield potential of 175-200 q/ha.

RAKSHA SUGANDH (DARL-52)

This variety has been developed at DARL, Pithoragarh and identified by AICRP-VC in 2002. Plants are about 70-80 cm tall. Bulbs are creamy white in colour with medium size each having about 20 cloves. The quantity of total soluble solids is fairly high in it amounting to 37.11%. It is suitable for processing with 1.5% mineral content. It has been recommeneded for the cultivation in J&K, H.P. and Uttaranchal and has yield potential of 100-150 q/ha in 210 days of crop duration.

Chapter 5

Cole Crops

CAULIFLOWER

Brassica oleracea var. *botrytis*
Family: Brassicaseae
Chromosome no.: 2n=18

India commands the largest area and production under cauliflower in the world and shares ~35% of total area and production. But in terms of productivity, it equals to the world average of 18.4 tonnes. Cauliflower is considered to be originated in the island of Cyprus from where it moved to Syria, Turkey, Italy, Spain and north western Europe. In India, it was introduced in 1822 from Kew Garden, London. In India, cauliflower is grown both in hills and plain mainly in states like Bihar, Uttar Pradesh, Orissa, West Bengal, Assam, Haryana, Maharastra, Himanchal Pradesh and Uttaranchal. Bihar, West Bengal and Uttar Pradesh are the leading producers of cauliflower in the country. Cauliflower is considered to be winter season vegetable but with the development of the tropical types, it is being cultivated throughout the year. Cauliflower is used in the preparations as mixed vegetable or curry. It is also used in the preparation of pickle along with other vegetables.

Based on temperature requirement for curd initiation and development, cauliflower is grouped into three maturity groups *viz.*, early, mid and late group. The sowing time of early maturity group starts from mid May followed by mid early (July end), mid late (August end) and late group (September to mid October). Well-drained, sandy loam to clay soil having pH of 6.0-7.0 and rich in organic matter is suitable. Cauliflower is a thermo-sensitive crop thus curd formation, bolting and vegetative growth is highly influenced by temperature. Therefore, cultivars are carefully selected according to the planting season. Seedlings become ready for transplanting in 4-6 weeks depending upon the weather conditions. Seedlings of early maturity group cultivar are usually transplanted on ridges. While late group cultivars may be transplanted on plain surface. To maintain proper soil moisture, irrigation is applied just after transplanting. The moisture stress during vegetative growth and at the time of curd initiation should be avoided. Early maturity group cultivars usually yield 10-12 t/ha, while it is 20-30 t/ha with late cauliflower.

Damping off is the most destructive disease of nursery, while *Rhizoctonia* rot and black rot are other major diseases. Among the insects, diamondback moth (DBM) and aphid cause severe damage.

Varieties

Most of the varieties of cauliflower have been developed using mass and recurrent selection, hybridization and synthesis. The improvement work on this crop is in progress at IARI (New Delhi), IARI Regional Station (Katrain), IIVR (Varanasi), and GBPUA&T (Pantnagar). Several varieties of different maturity groups have been developed and described in detail as under (for photographs see plate 11).

PUSA SNOWBALL-16

This late maturity group variety has been developed at IARI Regional Station, Katrain and notified by CVRC Notification nos. 716, 361(E), 440, 13 dated 20.02.1970, 30.06.1973, 21.07.1975, 19.12.1978, respectively. Curd is white, solid and self-blanched. This variety has yield potential of 150-200 q/ha.

EARLY KUNWARI

This variety has been developed from a selection at IARI, New Delhi and notified by CVRC Notification no. 1004 and 361(E) dated 23.03.1970 and 30.06.1973. Leaves are bluish green, and wavy with waxy bloom. Curds are semi-spherical with even surface. Curd availability period is mid September to mid October for May sown crop. It is ready for harvesting after 100-200 days of sowing. It has been recommended for the cultivation in states of Punjab, Haryana, H.P. and Delhi and has yield potential of 160-180 q/ha.

PUSA KATKI (PUSA KATKI-7)

This variety has been developed at New Delhi and notified by CVRC Notification no. 716, 440, 786, and 13 dated 20.02.1970, 21.08.1975, 02.02.1976 and 19.12.1978, respectively. It is one of the earliest variety, which gets matured in October-November and having medium plants, bluish green, waxy leaves. It is suitable for sowing in the middle of May. Late planting beyond middle of August does not yield good sized curds. It has yield potential of 130-140 q/ha.

HISAR-1

This variety has been developed at HAU, Hisar and notified by CVRC Notification no. 786 dated 02.02.1976. It bears medium to large size white heads. It is suitable for mid and late season cultivation under Haryana conditions and has yield potential of 250-275 q/ha.

PUSA DEEPALI

This variety has been developed at IARI, New Delhi and notified by CVRC Notification no. 1004 dated 23.03.1970. Plants are medium tall. Leaves are erect, short, green and waxy. Curds are compact, self-blanching, white and medium in size. Uniform curds are well protected by leaves and resulting no riceyness. Curds are ready in late October when the average temperature is around 20-25°C. It is recommended for sowing from May end to early June. It is recommended for general cultivation in Delhi and Punjab and has yield potential of 150-175 q/ha.

PUSA SNOWBALL-1

This variety is a derivative of a cross (EC-12013 × EC-12012) has been developed at IARI, Regional Station, Katrain and notified by CVRC Notification no. 1004 dated 23.03.1970. It is a late maturity group variety, suitable for cool season. The optimum average temperature for curd initiation and development is 10-16°C. It's sowing time in north India is from mid-September to the end of October. Leaves are straight, upright and inner leaves tightly cover the curd. Curds are very compact, medium in size and snow white. This variety has yield potential of 140-160 q/ha.

PUSA SNOWBALL-2

This variety has been developed at IARI, Regional Station, Katrain and notified by CVRC Notification no. 1004 dated 23.03.1970 for general cultivation throughout the country. It is suitable for late sown conditions. Its outer leaves are upright while inner leaves cover the curd initially. Curds remain white even on exposure to sun light. Curds become ready by the end of January or beginning of February. Seed yield of this variety is poor. It has yield potential of 170-190 q/ha.

PUSA SHUBRA

This variety has been developed at IARI, New Delhi and notified by CVRC Notification nos. 295(E), and 10(E) dated 09.04.1985, and 01.01.1988, respectively. Plants are erect with long stalk and bluish green leaves. Curd is compact and white each weighing 700-800 g. Plants are highly tolerant to riceyness and are resistant to black rot under field and artificial inoculation conditions. It takes 125-130 days for 50% curd formation and best temperature for curd is 12-16°C. The yield potential of this variety is 250-300 q/ha.

PUSA SNOWBALL (K-1)

This variety has been developed at IARI Regional Station, Katrain and notified by CVRC Notification no. 10(E) dated 01.01.1988. Plants are tolerant to black rot (*Xanthomonas campestris*). It bear best quality curds, which are snow white in colour and retain it even if the harvesting is delayed. Leaves are puckered, serrated and light green, harvesting; tolerant to black rot. It has been recommended for cultivation in U.P. and has yield potential of 200-225 q/ha.

PANT GOBHI-3

This variety has been developed at GBPUA&T, Pantnagar and notified by CVRC Notification no. 280(E) dated 13.04.1989. It is a synthetic variety combining eight inbred lines of katki group. Plants have relatively longer stem and sparse semi-erect leaves. Its curds are hemispherical, creamy white, medium compact, non-ricey and mature in 110 days. It has yield potential of 130-150 q/ha.

PUSA EARLY SYNTHETIC

This variety has been developed at IARI, New Delhi and notified by CVRC Notification no. 617(E) dated 17.08.1993. Plants are erect with bluish green leaves. Curds are small-medium, flat, creamy white and compact. It is suitable for early planting and resistant to riceyness. This variety has been recommended for the cultivation in Punjab, U.P., Bihar, Karnataka, Tamil Nadu and Kerala and has yield potential of 120-150 q/ha.

PUSA HYBRID-2

This Hybrid has been developed at IARI, New Delhi and notified by CVRC Notification no. 636(E) dated 02.09.1994. It is semi-erect with bluish-green upright leaves. Curds are creamy-white, highly compact with average curd weight of 907 g. It matures from mid November to mid December in the North Indian Plains. Plants are resistant to downy mildew. It has been recommended for cultivation in Humid Bengal-Assam basin and Sub-Humid Satluj Ganga Alluvial plains and has yield potential of 250-270 q/ha.

PANT GOBHI-4 (235-S)

This variety has been developed at GBPUA&T, Pantnagar and notified by CVRC Notification no. 408(E) dated 04.05.1995. It is a November maturity group variety derived through simple recurrent selection from a local variety (Aghani). Plants have medium-long stem and sparse semi-erect leaves with 15 cm of stalk. Curds are semi-spherical, creamish white, medium, compact and non-ricey. Its takes about 115-

120 days for curd maturity after transplanting. It has been recommended for the cultivation in West Bengal, Assam, M.P. and Maharshtra and has yield potential of 150-170 q/ha.

KASHI KUNWARI (IVREC-2)

This early maturity group variety has been developed at IIVR, Varanasi and notified by CVRC Notification no. 597(E) dated 24.04.2006. Under Indian plain conditions, its suitable sowing time is end of June to July. It can tolerate high rainfall during its vegetative growth. Curds are semi-dome type, white compact, texture fine with curd weight of 300-450 g. It has been recommended for cultivation in Punjab, U.P., Bihar and Jharkhand and has yield potential of 17-180 q/ha.

PUSA SHARAD (SEL-309-1-2)

This variety has been developed at New Delhi and notified by CVRC Notification no. 642(E) dated 31.05.2004. Plants are semi erect and open type with small stalk. Leaves are glabrous, bluish-green, long petiolate, oblong with narrow apex, wavy margin, prominent mid-rib and ear-like lobes at the base of lamina. Its marketable curd becomes ready after 85 days of transplanting. Curds are white, knobby, very compact, semi-dome shaped and about 750-1000 g in weight. It has been recommended for cultivation in Uttar Pradesh, Punjab, Haryana, Bihar, Rajasthan and West Bengal and has yield potential of 260-280 q/ha.

PUSA MEGHNA (DC-98-2)

This variety has been developed at IARI, New Delhi and notified by CVRC Notification no. 642(E) dated 31.05.2004. Curds are retentive white, compact and medium size (400 g). Curd becomes ready for harvest after 95 days of seed transplaning. It has yield potential of 125-150 q/ha.

IMPROVED JAPANESE

This November maturity group variety has been introduced from Israel at IARI, New Delhi and identified for release through AICRP-VC in 1975. It is erect with bluish green leaves. Curds are white, and compact. It is susceptible to heat and curd availability period is late

November to December (16-20°C) from late July to early August sown crop. This variety has been recommended for cultivation in Indian plains and has yield potential of 110-125 q/ha.

PUSA SNOWBALL

Pusa Snowball is a January maturity group variety has been derived through selection from a cross (EC-12013 × EC-12012) at IARI, Katrain and identified for release through AICRP-VC in 1975. Leaves are straight and upright. Curds are very compact, medium, snow white and tightly covered under inner leaves. Curd availability period is January to February from mid September to end October sown crop. It is suitable only for cool season cultivation. It has been recommended for cultivation throughout India and has yield potential of 170-200 q/ha.

PUSA KARTIK SANKAR (DCH-541)

This hybrid has been developed at IARI, New Delhi and identified for release through AICRP-VC in 2002. Temperature requirement for curd formation is 22-25°C. Maturity is 96 days, which is 2 weeks earlier than Pusa Early Synthetic. Curds are compact, white, medium sized and each weighing about 475 g. It is resistant to downy mildew. It has been recommended for cultivation in West Bengal, Assam, Punjab, U.P., Bihar and Jharkhand and has yield potential of 145-160 q/ha.

SUMMER KING

This hybrid has been developed at Sungro Seeds Pvt Ltd and identified for release through AICRP-VC in 2004. It produces whitish round curd and ready for harvesting with 65 days after transplanting. It can be grown in J&K, H.P., Uttaranchal, Punjab, U.P., Bihar, and Jharkhand and has yield potential of 200-225 q/ha.

CABBAGE

Brassica oleracea var. *capitata*
Family: Brassicaceae
Chromosome no.: 2n=18

Cabbage holds a coveted position among the vegetables due to its area under cultivation, production and productivity. Cabbage is a native of southern Europe and the Mediterranean region. The original farm was non-heading type. The head cabbage was referred much later in about 16th century. Cabbage probably originated from a wild ancestor *Brassica oleracea* var. *sylvestris*. Several wild forms of *Brassica* species are found along the coast of the Mediterranean sea and from there spread over to European countries. The wild form of cabbage still exists in Western Europe and South Africa. In India, it is introduced during Mughal period. In India, it occupies second place after cauliflower. Among the vegetables, cabbage having maximum share under hybrid seed and now can be grown round the year due to the development of tropical types. Cabbage is characterized by its swollen heads, which is formed by thickening of edible bud with overlapping leaves manifesting a large head. Cabbage cultivars can be classified into three broad groups, *viz*., (i) white cabbage, (ii) red cabbage, and (iii) savoy cabbage. White cabbage is of commercial importance in India. Red and Savoy cabbage are grown on very limited scale.

Cabbage is a good source of vitamins A, B_1 and B_2 and an excellent source of vitamin C. The 100 g edible portion of cabbage has 92.1% moisture, 25.1 cal food energy, 1.7 g protein, 0.2 g fat, 5.3 g total carbohydrates and 0.9 g fibre. Cabbage is cooling in effect and helps in preventing constipation, increases appetite, speeds up digestion, and is very useful for diabetes patient. It is used as salad, boiled vegetable, cooked in curries, pickling as well as processed as dehydrated vegetable. Cabbage is gaining popularity in Indian diet and now it is at par in consumption with cauliflower.

Cabbage is available upto April in northern plains. The yield level is very low that of compared to developed countries. The reason for low yield are lack of suitable cultivars and poor seed quality. Traditional cultivars perform poorly both in terms of head quality and yield. Southern and south eastern parts of India are the best suited areas for commercial cultivation of cabbage. Uttar Pradesh, Orissa, Bihar, Karnataka, West Bengal and Assam are other important cabbage growing states in India.

Cabbage can be grown in almost all types of soil but for late crop clay loam with pH of 5.5-6.5 is best. Relatively cool and moist climate is the best for cabbage growth. Cabbage is biennial having vegetative and productive phases in different seasons. It is requires low temperature (chilling requirement) for bolting leading to reproductive phase. Therefore, seed production of cabbage is possible only in temperate climates. Cabbage can be grown throughout the year in mid hills and southern hills, while in plains and foot hills zones of India, it is cultivated from October to January. The practice of nursery raising is similar to cauliflower. Late crops require 375 g/ha while early crop requires 500 g/ha of seed. The 4-6 weeks old or 4-6 true leaf stage seedlings are transplanted.

Heads become ready for harvesting when they are firm and tender. Harvesting of cabbage head should be done at right stage to avoid cracking and rotting of heads. The major diseases in cabbage cultivation are damping off, black rot and soft rot, whereas, major insects are caterpillar and aphids.

Varieties

In cabbage improvement programme, main emphasis is given for development of high yielding hybrids for both temperate and tropical conditions. The research programmes in India on cabbage are being carried out at IARI Regional Station (Katrain), HPKV (Palampur), and YSPUH&F (Solan). Besides, heterosis breeding programmes are carried out mainly by private seed companies. Several varieties of different maturity groups have been developed as detailed below (for photographs see plate 12).

GOLDEN ACRE

This variety has been developed through selection from Copenhagen Market at IARI, Regional Station, Katrain and notified by CVRC Notification nos. 716, 361(E), 440, 13, 786 and 540(E) dated 20.02.1970, 30.06.1973, 21.08.1975, 19.12.1978, 02.02.1976 and 24.07.1985, respectively. It takes about 60-65 days from transplanting to head formation. It has fewer outer leaves, which are cup shaped and arranged in two whorls. Heads are solid, short core and weighing of about 1.0-1.5 kg. It should be harvested immediately after head formation otherwise cracking may take place. It is highly suitable for spring and summer season cultivation in the hills and has yield potential of 175-200 q/ha.

PUSA DRUM HEAD

This variety has been developed at IARI, Regional Station, Katrain and notified by CVRC Notification no. 716, 540(E) dated 20.02.1970, 24.07.1985, respectively. It has wider frame with 20-25 light green outer leaves with prominent midribs and venation. Heads are solid and flat weighing about 3 to 4 kg. It takes 80-90 days from transplanting to head formation and is probably the earliest variety among the Drum Head Group. It is field resistant to black leg and has yield potential of 220-210 q/ha.

PRIDE OF INDIA

This variety is an introduction, recommended by YSPUH&F, Solan and notified by CVRC Notification nos. 361(E), 440, and 540(E) dated 30.06.1973, 21.08.1975 and 24.07.1985, respectively. The plant type of this variety is similar to 'Golden Acre' but it is about a week late in maturity. It has bigger heads each weighing 1.5 to 2.0 kg. It has yield potential of 215-240 q/ha.

EARLY DRUMHEAD

This variety has been developed at IARI, Regional Station, Katrain and notified by CVRC Notification no. 13 dated 19.12.1978. It is an early variety having light green foliage. Leaves are medium to large, curving inward and enclosing the head loosely. Leaves forming the head fold over each other at the centr and heads are flat. It has yield potential of 180-200 q/ha.

COPENHAGEN MARKET

This variety is an introduction, recommended by IARI, Regional Station, Katrain and notified by CVRC Notification no. 13 dated 19.12.1978. It is still being grown but on a limited area. This is an introduction, which has been replaced by Golden Acre and Pride of India because of their earliness and smaller head sizes. It has yield potential of 175-200 q/ha.

PUSA MUKTA (SEL-4)

This variety has been developed through hybridization (EC-24855 ×

EC-10109) followed by pedigree selection at IARI, Regional Station, Katrain and notified by CVRC Notification no. 280(E) dated 13.04.1988. It has short stalk, medium frame, and slightly bigger leaves than Golden Acre and wavy puckered at the margins. Heads are compact, slightly flatish-round with loose wrapper leaf at the top each weighing 1.5-2.0 kg. It is about a week late than Golden Acre in maturity and average head weight is 1.5-2.0 kg. It is resistant to black rot (*X. campestris*) and has been specially identified for the areas, where this disease is a severe problem. It has yield potential of 200-225 q/ha.

PALAMPUR GREEN

This variety has been developed at HPKV, Palampur and notified by CVRC Notification no. 115(E) dated 10.02.1996. Leaves are broad, smooth and dark green with prominent white and tender petiole. This variety has yield potential of 380-400 q/ha.

KINNER RED

This variety has been developed at IARI Regional Station, Katrain and notified by CVRC Notification no. 1052(E) dated 26.10.1999. Head round at top and oval at the base purple colour with red pigmentation and leaves upright clasping around the head. It has yield potential of 250-275 q/ha.

SRI GANESH GOL

The hybrid has been developed at Maharashtra Hybrid Seeds Co. Ltd., Jalna and identified for release through AICRP-VC in 1992. Plant vigorous, heads round, compact, attractive, bluish green and become ready for harvest in 90 to 95 days after sowing. It is resistance to yellowing and has good transport quality. It has been recommended for cultivation in Andhra Pradesh, Chhattisgarh and Orissa and has yield potential of 850 to 875 q/ha.

SUVARNA (BSS-32)

The hybrid has been developed at Beejo Sheetal Seeds Private Limited, Jalna and identified for release through AICRP-VC in 1995.

Heads are round with average weight of 3 kg. It has strong and smooth outer leaves. It is recommeneded for cultivation in the states of Madhya Pradesh and Maharashtra and has yield potential of 275-300 q/ha in 110 days of crop duration.

QUISTO

This variety has been developed at Novartis Seeds Private Limited, and identified for release through AICRP-VC in 1998. It is a high yielding hybrid and has ability to withstand over severe hot conditions. Heads can stand in the field conditions up to 70 days. Heads are compact round and very solid with dark blackish colour. It has been recommeneded for cultivation in Punjab, U.P., Bihar and Jharkhand and has yield potential of 300-325 q/ha.

BROCCOLI

Brassica oleracea var. *italica*
Family: Brassicaceae
Chromosome no.: 2n=18

Broccoli, commonly known as green cauliflower in India is a recent introduction from USA and Europe. The name 'broccoli' is derived from an Italian word "*brachium*" means an arm band. The term broccoli is also applied to young floral shoots of several *Brassica* crops, like, cabbage and turnip, but usually refers to heading and sprouting forms belonging to the *Brassica oleracea* L. italica group. The origin of broccoli and cauliflower seems to be common. However, it has been suggested that broccoli might have originated as an abnormality by crossing between early forms of branching shrub kale and other botanical varieties. Eastern Mediterranean areas and Italy are considered as centres of its diversification.

It is easily distinguished from cauliflower by having a head composed of differentiated flower buds, rather than curd. The two main forms are commonly cultivated, *i.e.*, sprouting broccoli and cabbage. The head is generally green or purple in colour, which is more nutritive than white colored broccoli. After cutting main head, smaller heads, which are called spear are developed in the axils of the leaves and harvested before becoming loose. Both main head and spears are used in fresh form. The edible portion is inflorescence consisting of immature fully differentiated flower buds and tender portion of the upper stem. Flower buds at the terminus of each branch of the inflorescence collectively produce a semi compact hemispheric head. The head diameter may range from 15 to 30 cm. Green/purple bud clusters are surrounded with leaves but leaves do not cover the head. The flowering stems are taller, 50-90 cm and have longer internodes than either cauliflower or cabbage. It is gaining popularity due to its nutritional and medicinal importance. Broccoli can be grown during winter season in India.

In India, it is grown as a winter crop in plains, whereas in hills it is grown as a spring-summer crop. It is highly sensitive to higher temperatures as bud clusters of head loose quickly. Optimum temperature for head development is 12-18°C. In North Indian plains, sowing time in nursery bed is mid August to mid September. Sowing in hills is done in July-August, whereas in high hills, it is done in March-April. About 300-400 g seeds are required for planting one hactare of land. Seedlings become ready 30-35 days after sowing. Delay in transplanting results poor root growth and undeveloped head (small in size). Varieties of early group mature in 40-45 days, mid group in 60-100 days and late group in >100 days of transplanting. Moisture stress during early vegetative growth and head development satage reduces the productivity. Irrigation at an interval of 7-8 days are required for good quality heads and more yield. When the head has attained the proper size, harvesting should be done immediate to avoid opening of buds (beads) and head loosing. The problem of diseases and pest is comparatively lesser than the cauliflower.

Varieties

In cabbage improvement programme, main emphasis is given for development of high yielding hybrids for both temperate and tropical type. The genetic improvement programmes on broccoli is being carried out at IARI Regional Station (Katrain), HPKV (Palampur), and PAU (Ludhiana). Besides, heterosis breeding programmes are carried out mainly by private seed companies. Several varieties of different maturity groups have been developed and described in detail as under (for photographs see plate 13).

PALAM HARITIKA

This variety has been developed at HPKV, Palampur and notified by CVRC Notification no. 664(E) dated 11.05.2005. It is a sprouting broccoli with dark green upright leaves having purple reddish tinge. The heads attain marketable stage in about 145-150 days. This variety has yield potential of 230-250 q/ha.

PALAM VICHITRA

This variety has been developed at HPKV, Palampur and notified by CVRC Notification no. 664(E) dated 11.05.2005. It is a heading broccoli, medium-sized heads with open dark green leaves. Purple tinge is prominent on stem at seedling stage and on leaf margins at full growing stage. The head is purple and compact. The variety is rich in nutrients, especially vitamins and minerals with high cosmetic appeal. The heads attain marketable size in about 115-120 days. It has yield potential of 230-260 q/ha.

PALAM KANCHAN

This variety has been developed at HPKV, Palampur and notified by CVRC Notification no. 664(E) dated 11.05.2005. It is a heading broccoli with long, broad, bluish green upright leaves having prominent white midrib and veins. The head is large in size, compact, attractive yellowish green in colour. It is rich in vitamin A. The heads attain marketable size in about 140-145 days. It has yield potential of 250-275 q/ha.

KNOL-KHOL

Brassica oleracea var. *gongylodes*
Family: Brassicaceae
Chromosome no.: 2n=18

Among the cole crops, Knol-Khol is considered as a minor crop since it is not grown on a commercial-scale anywhere in the world. Knol-khol is originated in North Eastern Europe and wild cabbage (*Brassica oleraceae* var. *slyvestris*) is considerd to be its progenitor. In India, it is popular in Kashmir, West Bengal, Maharastra, Assam, Uttar Pradesh, Punjab and some parts of south India. The crop is grown for above ground swollen stem known as 'Knob' (turnip-like shape). Its flavour is similar to that of turnip. It can be eaten as raw or cooked. It is used as salad, in curries, stews or boiled as pickle. The leaves of young plants may be used like green leafy vegetables. Knol-khol is a rich source of Ca, Mg, K, P, B, Na and S.

Knol-khol can be grown in all type of soil rich in organic matter. Cool climate (15-25°C temperature) is highly preferable for its growth and development. September-November is suitable time of seed sowing in nursery bed. In hilly areas, it can be sown in March-April. The seedlings are ready for transplanting after 30-35 days of sowing. About 800-1000 g seeds are required for raising seedlings to be planted in one hectare. Proper moisture is maintained during its early growth to avoid poor growth and fibrous knobs. The edible maturity of knobs depends upon the varieties. Generally, the knob size at the time of maturity attains 6-8 cm diameter is early varieties and 10-12 cm diameter in late maturity varieties.

Damping off is primary disease of nursery and causes the loss at both pre and post-emergence stage. Besides, black spot, downy mildew, *Rhizoctonia* and black rot are other diseases causing yield loss.

Varieties

The genetic improvement work in knol-khol is restricted to introduction and evaluation of varieties obtained from European countries. The introduced varieties are multiplied and maintained for purity.

LARGE GREEN

This variety has been developed at IARI, Regional Station, Katrain and notified by CVRC Notification no. 440, dated 21.08.1975. The knobs are early with small tops tender, and flavoured. This variety has the yield potential of 225-250 q/ha.

WHITE VIENNA

This variety has been developed at IARI, Regional Station, Katrain and notified by CVRC Notification no. 440, dated 21.08.1975. The knobs are white, medium sized, tender, oval-shaped with distinct small tops. It has the yield potential of 180-200 q/ha.

PALAM TENDERKNOB

This variety has been developed at HPKV, Palampur and notified by CVRC Notification no. 664(E) dated 11.05.2005. It is characterized by small green foliage. The knobs are round, flat, stringless and fleshy. Knobs attain marketable size about a week earlier than White Vienna. It has yield potential of 250-275 q/ha.

Chapter 6

Root Vegetables

CARROT

Daucus carota
Family: Umbellifereae
Chromosome no.: 2n=18

Carrot is an important root vegetable grown in spring, summer and autumn seasons in temperate and during winter in tropical and subtropical climates. Carrot is an important source of carotene. South western Asia, especially Afghanistan is considered to be the centre of origin. It is speculated that mutation and selection was more responsible to the development of the cultivated carrot than by hybridization. At present two main carrot types, *i.e.,* Anthocyanin or Asiatic types and carotene or orange colored Temperate/European types are grown on commercial scale.

The inflorescence of carrot is a compound umbel. A carrot plant usually takes 4-6 weeks for complete blooming. Closed anther mutants with unrolled stamen filaments have been observed. This type of functional male sterility is useful in commercial hybrid seed production. In some varieties, protandry is present where pollens mature 3-4 days before stigma receptivity, which is also responsible for cross-pollination.

Carrot develops a deep, extensive root system going as deep as 75-90 cm. The long thin taproot rapidly elongates and achieves its potential length in 12-24 days after emergence. With enlargement of edible taproot, a large number of absorbing lateral roots also develop. Botanically the fleshy taproot is designated as conical root. Primarily it is consists of phloem (cortex) and xylem (core). Good quality roots have maximum of cortex and minimum of core. Self-coloured core is desirable.

Well-drained loamy soil is ideal for proper growth and development of carrot roots. Soil pH of >7.1 or <6.6 leads to reduction in yield. An average temperature of 18-22°C is the best for root development. Crop can be grown directly from seeds, which may be sown on flat beds or ridges. The seed requirement for one hectare area is 5-6 kg. Water soaking of seeds for 24 hours before sowing facilitates good germination. Frequent irrigation, to maintain the optimum moisture level at root development, prevents secondary root development. *Cercospora* leaf blight and soft rot are the major diseases of carrot, whereas, carrot rust fly is the main insect damaging the roots. Several physiological parameters, *i.e.*, forking (due to soil), splitting (due to sudden change in soil moisture), and cavity spot (calcium deficiency) contribute to unmarketable yield. Right stage of maturity can be judged by measuring the diameter (2-4 cm) at upper end of roots. Before harvesting a light irrigation facilitates the pulling of roots without any damage. The crop improvement programmes mainly aim to develop varieties with high carotene content and yield. Thick fleshy roots with thin self-colored core along with resistance to non-branching and non-cracking are another target of carrot improvement.

Varieties

Mass and pedigree selection are common breeding methods used for population improvement. Hybrids are being developed by utilizing cytoplasmic male sterile lines. The improvement of carrot in India is being carried out by IARI (New Delhi), IARI Regional Station (Katrain), IIVR (Varanasi), SKUA&T, Srinagar, and PAU (Ludhiana). Several varieties of different maturity groups have been developed and described in detail as under (for photographs see plate 14).

NUMBER-29

This is an Asiatic type variety developed through selection from local material at PAU, Ludhiana and notified by CVRC Notification no. 361(E) dated 30.06.1973. Roots are long, tapering, light red and thin. It has yield potential of 275-325 q/ha.

NANTES

This European variety is an introduction, recommended by IARI, Katrain and notified by CVRC Notification nos. 361(E), 440, and 540(E) dated 30.06.1973, 21.08.1975, and 24.07.1985. It can be grown in the plains of India for root production but not for seed production. Its seeds are produced in only hilly regions. Roots are half-long, slim, bell shaped, cylindrical with a small thin tail, deep orange-red cortex and core. It ranks first in quality but has a weak, brittle top that makes it difficult for pulling. Keeping quality is low due to thin skin and fine texture. It is suitable for cultivation in winter season and has yield potential of 325-350 q/ha.

EARLY NANTES

This European variety is an introduction, recommended by IARI, Katrain and notified by CVRC Notification nos. 13, and 540(E) dated 19.12.1978, and 24.07.1985. It has almost cylindrical roots terminating abruptly in small thin tail, it is 12-15 cm long with fine texture. Flesh is orange with self-coloured core. It takes 90-100 days for root formation and 310-330 days for seed crop. This variety has yield potential of 300-350 q/ha.

SEL-233

This variety has been developed through hybridization (Nantes × No.-29) followed by pedegree selection at PAU, Ludhiana, and notified by CVRC Notification no. 13 dated 19.12.1978. It has small top and dark green leaves if 30-35 cm. Roots are ~15 cm long, smooth, semi cylindrical, orange with light orange core show less cracking and forking. Harvesting can be delayed without bolting and impairing its edible qualities. It takes 90-100 days for root formation and has yield potential of 250-275 q/ha.

PUSA MEGHALI (SELECTION-5)

This variety has been developed through hybridization (Pusa Kesar × Nantes) followed by pedigree selection at IARI, Regional Station, Katrain and notified by CVRC Notification no. 636(E) dated 02.09.1994. It has short tops with smooth roots, orange flesh, self-coloured core and stumpy to slightly tapering roots. It is suitable for sowing as early (August-September) as well as late (October-November). It is preferred for early sowing and can be marketed in 110-120 days. It is recommeneded for the cultivation in Madhya Pradesh and Maharashtra and has yield potential of 250-280 q/ha.

PUSA YAMDAGNI (SELECTION-5)

This hybrid has been developed through heterosis breeding (EC-9981 × Nantes) at IARI, Regional Station, Katrain and notified by CVRC Notification no. 10(E) dated 01.01.1988. It combines the earliness of EC-9981 and self-coloured core character of Nantes. Roots are ~15 cm long, orange, and slightly tapering stumpy to semi-stumpy ending. Tops are medium-sized and quick growing as compared to other temperate varieties. It is high yielder and richer in carotene content. It is recommeneded for cultivation in Madhya Pradesh and Maharashtra and has yield potential of 225-260q/ha.

OOTY-1

Ooty-1 is a selection from half-sib progeny of local type (DC-3), developed at TNAU, Coimbatore and notified by CVRC Notification no. 1052(E) dated 26.10.1999. Roots are long, slightly tapering with attractive deep orange colour. Central core is thin, fleshy and palatable. This is a medium maturing (100-110 days) variety suitable for round the year cultivation in hilly areas at altitude above 1800 m. The roots are rich in beta-carotene. It is free from pre-mature bolting and resistant to powdery mildew, leaf spot and drought and has yield potential of 460-490 q/ha.

SEL-21

This variety has been developed at PAU, Ludhiana and notified by CVRC Notification no. 165(E) dated 06.03.1987. Roots are medium long, red with self-coloured core. This variety has yield potential of 180-200 q/ha.

HISAR SELECTION-1

This variety has been developed at HAU, Hisar and notified by CVRC Notification no. 597(E) dated 25.04.2006. It is high yielding with tender white long roots. This variety has yield potential of 300-310 q/ha.

SHALIMAR CARROT-1 (SKAUC-50)

This temperate type variety has been developed through selection from a local population at SKUA&T, Shrinagar and notified by CVRC Notification no. 2035(E) dated 28.11.2006. It has dark green foliage, early maturing with high percentage of uniform cylindrical orange blunt roots with small self-coloured core. Roots are tender, sweet and fine textured, almost free from cracking and forking, resistant to frost and early bolting and rich in carotenes (56.1 mg/100g), dry matter (12.4%) and TSS (10.50%). It is recommeneded for cultivation in J&K, H.P. and Uttaranchal and has yield potential of 225-240 q/ha.

HYBRID NUMBER-1

This hybrid has been developed through heterosis breeding at Maharashtra Hybrid Seeds Company, Jalna and identified for release through AICRP-VC in 1992. Its roots are early maturing, uniform deep orange and 17-20 cm in length. This carrot hybrid is suitable for processing and fresh marketing. This hybrid has been recommended for cultivation in J&K, H.P., Uttaranchal, Karnatka, Tamil Nadu and Kerala and has yield potential of 400-450 q/ha.

RADISH

Raphnus sativus
Family: Brassicaceae
Chromosome no.: 2n=18

Radish is an important root crop of temperate and tropical countries. Roots and leaves and consumed as salad and cooked vegetable. The juice of fresh leaves is useful as diuretic and laxative. The origin of the radish is Europe. The crop is annual for root production, whereas it takes 2 years for seed production. It is highly cross-pollinated and predominantly cool season crop however Asiatic types can be grown at higher temperatures. September to October is the main growing season. Now Asiatic radish varieties are available around the year.

In Northern hills, seeds are usually sown from March to the beginning of November. Whereas in plains, August-January is the best sowing time of Asiatic and September-March for European types. In South Indian hills and plains, the best sowing time is April-June and October-December, respectively. Light soil is the best for radish cultivation. Seed requirement for one hectare is about 9-12 kg/ha. Sowing of seeds should be done on ridges followed by light irrigation and then weekly irrigation is suitable to get the quality roots. Once seedling are establishes, thinning should be done to maintain the proper distance. Shallow weeding helps proper root development. Edible maturity of roots depends upon the variety. The maturity of European cultivars is early (25-30 days after sowing) as compared to Asiatic types. Delayed ripening causes bitterness and pithiness of roots. A light irrigation before 2 days of harvest of roots facilitate easy harvest. Damping off and *Alternaria* blight are main diseases of radish, while aphid is a more serious pest. Proper plant protection measures should be adopted to manage the diseases and pest. Hollow root wart and pithiness are physiological disorder of radish leading to production of unmarketable root.

Varieties

Most of the varieties of radish have been developed through introduction, mass selection or through single-, double-, or a three-way cross. The improvement of radish in India is being carried out at IARI (New Delhi), IARI-Regional Station (Katrain), IIVR (Varanasi), IIHR (Bangalore), CSAU&T, Kanpur, TNAU, Coimbatore, and PAU (Ludhiana). Several varieties of different maturity groups have been developed and described in detail as under (for photographs see plate 15).

JAPANESE WHITE

This variety is an introduction (Shiroaguri-Kyo) from Japan, recommended by IARI, New Delhi, and notified by CVRC Notification nos. 361(E), and 440 dated 30.06.1973, and 21.08.1975. Roots are 25-30 cm long, 5 cm in diameter, cylindrical and blunt at the tip. Skin is pure white and smooth. Flesh is snow white, crisp, solid and mild pungent. Tops are medium large with deeply cut leaves. It is suitable for growing between October-December in the plains and July-September in the hills. It matures in 45 days and has yield potential of 175-200 q/ha.

WHITE ICICLE

This variety is an introduction, recommended by IARI, New Delhi and notified by CVRC Notification no. 361(E) dated 30.06.1973. It is a medium-short European variety that matures in 30 days. The skin is pure white, thin and tender, whereas the flesh is white icy, crisp, juicy, mild and of sweet flavour with enough pungency to appeal the appetite. Roots are solid, Icicle-shaped straight and tapered. Delay in harvesting leads to pithiness, suitable for October and November sowing in the plains. It has yield potential of 225-250 q/ha.

CHINESE PINK

This variety is an introduction, recommended by YSPUH&F, Solan, and notified by CVRC Notification no. 441(E) dated 21.08.1975. Roots are 12-15 cm long, semi stumpy to stumpy roots, pink with white colour towards the tip and takes about 50-55 days from sowing to root formation. It is a good cultivar for the hills but does well also in plains. It requires low temperature for seed production. It has yield potential of 250-275 q/ha.

NADAUNI

This variety has been developed from a local germplasm and notified by CVRC Notification no. 441(E) dated 21.08.1975. Roots are long, tapering, smooth and light pink in colour. The tops are dark green with cut leaves. This variety has yield potential of 175-225 q/ha.

PUNJAB SAFED

This variety has been developed by hybridization (White-5 × Japanese White) followed by selection at PAU, Ludhiana, and notified by CVRC Notification no. 13 dated 19.12.1978. Roots are white, tapering smooth, mild in taste, free of forking, 30-40 cm in length and 3.5 cm thick. It is quick growing types and roots remain edible for 10 days after attaining full size. It takes 50-60 days for root formation and has yield potential of 230-260 q/ha.

KALIANPUR NO-9

This variety is a selection from local material at CSAUA&T, Kanpur, and notified by CVRC Notification no. 2(E) dated 03.01.1983. It has green stem and foliage with smooth heavy top. Leaves are long, broad and less-lobed. Roots are 22-23 cm long, smooth, crispy, and white with green shoulder, thick and tapering. It is free from mustard sawfly, aphids and white rust and has yield potential of 250-300 q/ha.

CO-1

This variety has been developed through selection (RS-44-2-1) at TNAU, Coimbatore, and notified by CVRC Notification no. 596(E) dated 13.08.1984. The roots are milky white, long (22 cm), thick (12.5 cm girth) and weighing 220 g. Roots contain 10.9 mg vitamin C and 1.2% crude fibre in 100 g of edible roots. It bolts and sets seeds in the plains. Roots have better consumer preference. Top to root ratio is 0.90. It is suitable for growing round the year under different cropping systems and has yield potential of 200-250q/ha in 40 to 45 days of crop duration.

PUSA CHETKI

This variety is a selection from a material collected from Denmark, developed at IARI, New Delhi, and notified by CVRC Notification no. 10(E) dated 01.01.1988. Its roots are medium large and stumpy, pure white, tender and smooth and mildly pungent. It becomes ready for harvesting in 40-50 days after sowing. Being tolerate to high temperatures it is suitable for sowing from the mid of March to the mid of August. It shows profuse seed set in the plains. It may be grown as a summer and a monsoon crop from April-September and has yield potential of 200-250q/ha.

PUNJAB AGETI

This variety has been developed at PAU, Ludhiana, and notified by CVRC Notification no. 10(E) dated 01.01.1988. Its roots are long, thin with lower half white and upper part red. It is suitable for April to August sowing and has yield potential of 300-325 q/ha.

ARKA NISHANT

This Asiatic variety is an introduction from Singapore, recommended by IARI, New Delhi, and notified by CVRC Notification no. 527(E) dated 16.08.1991. Its roots are medium-sized (25 cm × 3-4 cm), marble white, crisp texture, pleasant aroma, free from early bolting, pithiness, splitting and forking and each root weighing 300-400 g. Root and shoot length ratio is 1:1 with mild pungency. It gets matures within 45-55 days after sowing and has yield potential of 200-300 q/ha.

KASHI SWETA (IIVR-1)

This variety has been developed through selection from Japanese germplasm, at IIVR, Varanasi, and notified by CVRC Notification no. 597(E) dated 25.4.06. It suitable for spring, summer, autumn and winter seasons sowing. Roots become ready for harvesting in 30-35 days after sowing. Roots are 25-30 cm long, 3.5-4 cm in diameter, straight, tapering with blunt tip. It has been recommeneded for cultivation in Uttar Pradesh and has yield potential of 480-500 q/ha.

KASHI HANS (IIVR-2)

This variety has been developed through selection at IIVR, Varanasi, and notified by CVRC Notification no. 597(E) dated 25.4.06. It is suitable for planting from September to February and Ist harvesting can be done within 40-50 days after sowing. Due to late pithiness, it can stand in the field upto 10-15 days after edible maturity. Its foliage is soft and smooth. Roots are straight and tapering of 35-40 cm long and 3.5-4.2 cm diameter. It is recommeneded for cultivation in Uttar Pradesh and has yield potential of 430-450q/ha.

BEET ROOT

Beta vulgaris var. *crassa*
Family: Chenopodiaceae
Chromosome no.: 2n=18

The name Beta came from Romans. This species is thought to be derived from *Beta Maritima*, known as sea beet, indigenous to southern Europe. It was taken to northern Europe by the invading armies, and it has been widely used throughout the Europe. Towards the end of the middle age, the garden beet with its thick cylindrical, or globular root had become an important vegetable in central Europe.

Beet root is commonly known as table beet. Based on the edible part, beets are generally classified into two major groups. The first one is mainly grown for its fleshy roots belonging to the *crassa* or garden beet group including yellow beets, common table beet, or red beet, while others are grown as leafy vegetable belong to the *cicla* group and include leafy beets and Swiss chard. The garden beet or table beet is also a popular root vegetable grown in home gardens as well as market gardens mainly for its fleshy enlarged roots. Beetroot is rich in proteins, carbohydrate, calcium, phosphorus, iron, vitamin A, and vitamin C.

Beet root is essentially a cool weather crop, and hence, grow best in winter in plains of India. It can also be grown in a bit worm climate, however, good quality roots rich in sugar and intense red colour are obtained in cool weather (18-21°C). It can be grown well in warm weather but the best quality beets are produced under cool weather conditions. Beet plants are very sensitive to low temperature. If they are exposed to relatively low temperature (4.5-10°C for 15 days or more), bolting is likely to occur before the roots reach marketable size. Good root colour is not directly related to high sugar content but low sugar content is associated with poor colour.

Beet root can be grown on a wide range of soils, however, deep and well-drained loams or sandy loams are considered the best, while the crop in heavy soils produces roots of asymmetrical shape. The crop is sensitive to acidic soil but thrives very well in alkaline soils with pH as high as 9-10.

In northern plains of India, usually, the seeds are sown in September to November, while in the southern plains its sowing is extended from July to November. In hills, seed is sown from first week of March to July end. Several successive sowings at an interval of 2 to 4 weeks ensure a continuous supply of marketable roots. Early sowing may produce roots with coarse and woody flesh with dull colour due to high temperature.

Seeds are usually sown directly in the field by sowing seed balls (seed mix with soil) that contain one to six seeds. The seed balls are sown at the rate of 7-9 kg/ha at 45-60 × 8-10 cm spacing. Seed treatment with Thiram or Captan at the rate of 2.5g/kg of seed gives better seedling emergence and control pre-emergence damping off disease. Seeds are generally soaked in water for about 12 hours before sowing to facilitate better germination in the field. Seeds are sown directly in the well-prepared fields on ridges at 45 to 60 cm apart, and are thinned later to space 8-10 cm within row. The seeds are soaked in water for about 12 hours before sowing to facilitate better germination in the field. Regular water supply is essential for both seed germination and high yield of good quality roots. The water requirement has been reported to be 300 mm supplied in five to six irrigations, however, when there is winter rains, three irrigations are sufficient for harvesting a good yield.

One or two earthing up is essential to prevent the exposure of roots to sunlight, which causes greening and lowers the quality of roots. In early stages, the crop growth is very slow, and the crop may not compete with weeds for nutrients, space, light, and moisture, hence, it is essential to remove the weeds. Thinning of young seedlings after germination provides the proper spacing between the plants. Usually, more than one seedling (4 to 6 seedlings) emerge from a seed ball, hence, thinning is essential to have roots of uniform shape and size. Hand thinning is done as and when the seedlings are large enough to handle. If poor germination is observed then the seedling uprooted during thinning may be transplanted, and soon after, a light irrigation may be applied. Thinning the seedlings to 3 cm produces roots of best quality.

Beet leaf miner or spinach leaf miner and beet webworm or leaf eating caterpillar are the major insects damaging the beet root crops. While, *Cercospora* leaf spot and *Sclerotium* root rot are the main diseases causing economical loss in the yield.

Varieties

Mass selection is generally adopted for improvement of turnip. The improvement work in India on this crop is being carried out at IARI Regional Station, Katrain and TNAU Research center at Ooty. Several varieties of different maturity groups have been developed and described in detail as under.

DETROIT DARK RED

This variety is an introduction, recommended by IARI, Regional Station, Katrain and notified by CVRC Notification nos. 13, and 540(E) dated 19.12.1978, and 24.07.1985. Top is small with dark green glossy foliage having maroon tinge. Midrib is thin from dorsal side but looks broad because of spreading maroon shade. Roots are smooth, uniform, and attractive with small collar and perfectly round with deep red skin. Flesh is blood red with light zoning, tender, round, fine grains and corrosive in taste when taken raw. It matures in 80-100 days and has yield potential of 250-300 q/ha.

CRIMSON GLOBE

This variety is an introduction, recommended by IARI, Regional Station, Katrain and notified by CVRC Notification nos. 13, and 540(E) dated 19.12.1978, and 24.07.1985. Top of this variety is medium to tall. Leaves are large, bright green, coppery shape with prominent venation, younger red veins. Roots are round to flatten round and medium red with indistinct zones. It is non-corrosive in taste when taken raw. It matures in 80-90 days and has yield potential of 200-250q/ha.

TURNIP

Brassica rapa
Family: Cruciferae
Chromosome no.: 2n=20

Turnip, grown for its enlarged root is an important winter vegetable in North India and sub-tropical regions of country. Mediterranean region is thought to be the primary centre of European types, while Eastern Afghanistan with adjoining areas of Pakistan is considered to be another centre of origin. The edible portion of turnip is swallow hypocotyls. Turnip is commonly used as salad or as cooked vegetable. Leaves are used as green vegetable. The shape and colour of underground portion vary with the cultivar. Generally, the root is flat globular, elongated and top shaped. Root colour of above ground portion is red, purple, white, yellow or green, while under ground potion may be white or yellow. Well-drained light soil with 5.5-6.8 pH and rich in organic matter is ideal for turnip cultivation. The cooler climate is better for quality produce, whereas high temperature adversely affects the quality. The optimum temperature for development of good quality roots is 10-13°C. Sowing in the plains of northern India starts from September and can be continued upto December. July to September period is suitable for sowing in hills. The seed requirement for one hectare is 3-4 kg. Seeds are sown on the ridges at a depth of 1.5-2.5 cm. Moisture in the field should be optimum at the time of sowing, otherwise pre-sowing irrigation is essential for better germination. To maintain optimum plant to plant spacing thinning is important. Harvesting at right stage of edible maturity is essential for getting quality roots. The ideal root diameter should be 5-7 cm, which varies as per cultivars. Delayed ripening of turnip often leads to bitterness and course texture. Light irrigation before, harvesting facilitates the harvesting.

Varieties

Mass selection is generally adopted for improvement of turnip. Hybridization approach may be followed by utilizing self-incompatible lines. The improvement work in India on turnip is being carried out at IARI Regional Station, Katrain. Several varieties of different maturity groups have been developed and described in detail as under.

SNOWBALL

This variety is an introduction, recommended by IARI, Regional Station, Katrain, and notified by CVRC Notification nos. 361(E), 440, and 13 dated 30.06.1973, 21.08.1975 and 19.12.1978, respectively. Bulbs become ready for harvest in 55 days. Leaves are small and light green. Roots are round and medium-sized with light yellow flesh. This variety has yield potential of 300-350 q/ha.

PURPLE TOP WHITE GLOBE

This variety is an introduction, recommended by IARI, Regional Station, Katrain, and notified by CVRC Notification nos. 440, and 13 dated 21.08.1975, and 19.12.1978, respectively. It is suitable for cool season cultivation. It is a high yielding and deep-rooted cultivar. Roots are nearly round, upper part purple and lower portion creamy. Flesh is white, firm, crisp and mildly sweet flavoured. The top is small, erect with cut leaves. It has yield potential of 275-325 q/ha.

Plate 1: Released varieties of tomato

Pusa Ruby

Kalyanpur Aungoorlata

Junagarh Ruby

Azad T-2

Pant Bahar

Sonali

Utkal Pallavi

Hisar Arun

Pusa Hybrid-2

Solan Sagun

Utkal Kumari

Mukthi

Swarna Baibhav

Kashi Vishesh

Kashi Amrit

Kashi Anupam

Kashi Sharad

Kashi Hemant

Arka Meghali

Swarna Naveen

Swarna Lalima

Azad T-6

GT-2

Punjab Kesari

Utkal Urbasi

BT-116-3-2

NDT-3

BT-20-2-1

Arka Vardhan

Avinash-2

KTH-2

KTH-1

Arka Alok

H-24

LE-415

Sun-496

Plate 2: Released varieties of brinjal

Pusa Purple Long

Bhagyamati

Azad Kranti

Azad B-1

Pant Samrat

Pant Rituraj

Punjab Barsati

Punjab Sadabahar

Aruna

Utkal Tarini

Pusa Hybrid-6

Hisar Shyamal

Hisar Pragati

Pusa Hybrid-5

GBH-1

NDBH-1

Pusa Upkar

Utkal Madhuri

Utkal Keshari

Sweta

Pusa Ankur

Azad Hybrid-1

Swarna Mani

Swarna Pratibha

Utkal Jyoti

Kashi Prakash

Kashi Taru

Swarna Shyamli

Swarna Shree

Kashi Sandesh

Arka Keshaw

Arka Nidhi

GBL-1

GBH-2

ARU-1

ARU-2C

Gulabi

KS-224

KS-331

Green Long

NDB-28-2

Pusa Shyamala

Pusa Hybrid-6

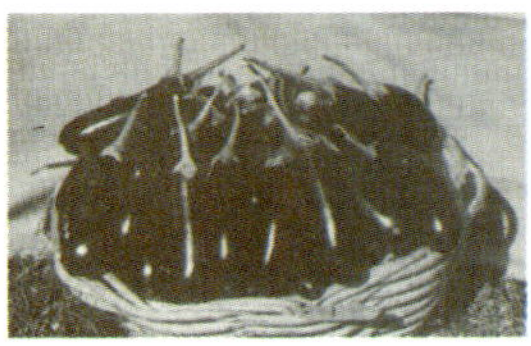

Pusa Hybrid-5

Phule Hybrid-2

Pusa Hybrid-9

SM-6-7

NDB-25

Plate 3: Released varieties of sweet pepper

California Wonder

Pusa Deepti

Nishat-1

Arka Basant

Arka Gaurav

Lario

DARL-202

Plate 4: Released varieties of chilli

Pusa Jwala

LCA-235

Jawahar Mirch-218

PKM-1

Jayanti

Kashmir Long-1

Utkal Ava

Prasanth

Arka Lohit

Azad Mirch-1

Kashi Surkh

GVC-111

GVC-121

G-4

LCA-206B

RHRC Cluster Ereet

ACS-2000-02

Sungrow-86-235

Utkal Rashmi

Kashi Anmol

Kashi Early

Plate 5: Released varieties of vegetable pea

VL-3

Pant Matar-2

VL-7

Ageta-6

VL-6

NDVP-8

Kashi Nandini

Kashi Udai

Kashi Mukti

Kashi Shakti

Arka Ajit

Pant Uphar

P-88

PH-1

DPP-9411

Plate 6: Released varieties of cowpea

Pusa Komal

Sel-263

Arka Suman

Arka Samuridhi

NDCP-13

Kashi Shyamal

Kashi Gauri

Kashi Kanchan

VR-5

Plate 7: Released varieties of French bean

VL Boni-1

Pant Anupama

Kashi Param

Arka Suvidha

Azad Rajmah-1

CH-812

Plate 8: Released varieties of Indian bean

Deepali

Swarna Utkrist

Wal Konkan-1

Plate 9: Released varieties of onion

Punjab Selection

P-48

Agrifound Dark Red

Pusa Madhvi

Arka Kalyan

VL Piaz-3

Agrifound Light Red

PBR-5

Arka Bindu

L-28

Arka Pitamber

S-131

Plate 10: Released varieties of garlic

Agrifound White

Yamuna Safed

Yamuna Safed-2

Yamuna Safed-3

VL Garlic-1

Yamuna Safed-4

DARL-52

Plate 11: Released varieties of cauliflower

Pusa Snowball K-1

Pant Gobhi-3

Pusa Early Synthetic

Pusa Hybrid-2

Pant Gobhi-4

Kashi Kunwari

Kashi Agahani

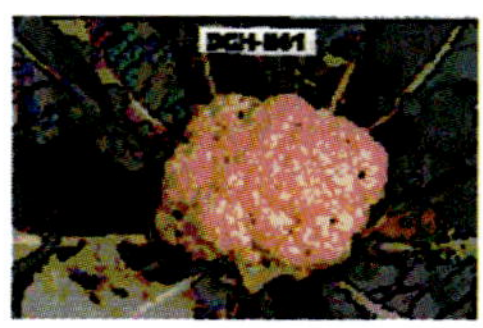

DCH-541

Summer King

Plate 12: Released varieties of cabbage

Golden Acre

Pusa Drum Head

Pusa Mukta

Sri Ganesh Gol

BSS-32

Quisto

Plate 13: Released varieties of broccoli

Palam Haritika

Palam Vachitra

Palam Kanchan

Plate 14: Released varieties of carrot

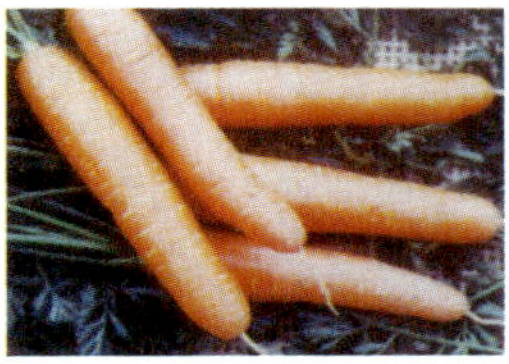

Pusa Yamdagni

Pusa Meghali

Shalimar Carrot-1

Ooty-1

Plate 15: Released varieties of radish

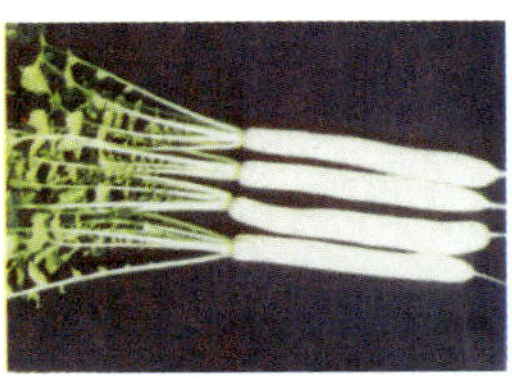

Japanese White

White Icicle

Kashi Sweta

Kashi Hans

Pusa Chetki

Hisar Sel-1

Plate 16: Released varieties of bottle gourd

Punjab Round

Kalyanpur Long Green

Arka Bahar

Rajendra Chamatkar

Pant Sankar Lauki-1

Narendra Sankar Lauki-4

Narendra Dharidar

Narendra Rashmi

Kashi Ganga

Kashi Bahar

Narendra Jyoti

Pant Lauki-3

Ajad Sankar-1

Pusa Naveen

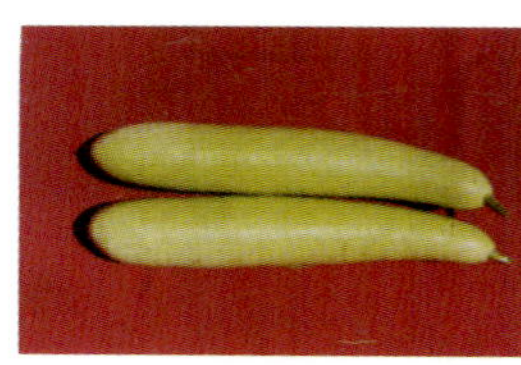
NDBG-132

Pusa Hybrid-3

Sel.P-6

CO-1

Plate 17: Released varieties of muskmelon

Hara Madhu

Durgapura Madhu

Punjab Sunehri

Gujrat Muskmelon-2

Arka Rajhans

Arka Jeet

Punjab Hybrid-1

Punjab Rasila

RM-43

MHY-3

RM-50

MHY-5

Gujrat Muskmelon-3

Kashi Madhu

NDM-15

Pusa Madhuras

DMDR-2

GM-1

Plate 18: Released varieties of bitter gourd

Kalyanpur Baramasi

Arka Harit

Kalyanpur Sona

Preethi

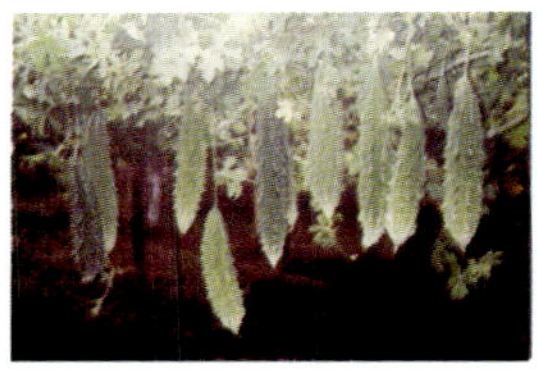

Phule Priyanka

Pant Karela-2

Pusa Hybrid-2

NBGH-167

Priya

Plate 19: Released varieties of cucumber

Kalyanpur Green

Swarna Poorna

Swarna Ageti

Pant Khira-1

Swarna Sheetal

Hybrid No. 1

Plate 20: Released varieties of pumpkin

Ambili

Pusa Vishwas

Arka Chandan

Kashi Harit

Sooraj

Narendra Agrim

Azad Pumpkin-1

Narendra Amrit

Saras

Narendra Abhushan

Arka Suryamukhi

CO-2

Plate 21: Released varieties of ridge gourd

PKM-1

Arka Sumeet

Arka Sujat

Swarna Manjari

Swarna Uphar

Deepthi

Plate 22: Released varieties of watermelon

Sugar Baby

Durgapura Meetha

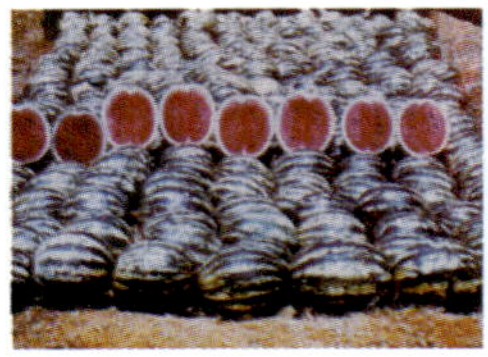

Arka Manik

Durgapura Lal

Plate 23: Released varieties of sponge gourd

Kalyanpur Hari Chikni

Azad Toria Chikni-1

Phule Prajakta

PSG-40

Plate 24: Released varieties of ash gourd

Pusa Ujjwal

Kashi Ujwal

Kashi Dhawal

PAG-72

Plate 25: Released varieties of pointed gourd

Rajendra Parwal-1

Rajendra Parwal-2

Swarna Rekha

Swarna Alaukik

Plate 26: Released varieties of longmelon

Arka Sheetal

Punjab Long Melon-1

Plate 27: Released varieties of roundmelon

Arka Tinda

Hisar Tinda

Plate 28: Released varieties of snake gourd

Konkan Shweta

Manushree

Plate 29: Released varieties of okra

Pusa Sawani

Pb-7

Utkal Gaurav

Varsa Uphar

Shitla Uphar

Shitla Jyoti

GO-2

Kashi Pragati

Kashi Vibhuti

Kashi Satdhari

Phule Utkarsha

Kashi Bhairav

Arka Anamika

Arka Abhay

Azad Bhindi-2

Kashi Leela

DVR-4

HRB-9-2

Chapter 7

Cucurbitaceous Vegetables

BOTTLE GOURD

Lagenaria siceraria
Family: Cucurbitaceae
Chromosome no.: 2n=22

Bottle gourd (syn. white flowered gourd, calabash gourd, Zucca melon) is commercially grown in all the states of India and available in the market throughout the year. Immature fruits are used in a number of ways. Bottle gourd is found in its wild form in South Africa and India, but probably it is indigenous to tropical Africa. It is good for people suffering from biliousness and indigestion. The fruit has a cooling effect, and beneficial for cardiac and diuretic problems. Its dry shells have been traditionally used as utensils and handicraft item in many parts of the world.

Plants are viny pubescent herb with stout five-angled stem. The stem is prostrate, profusely branched covers varying space depending upon weather conditions, soil nutrient, available moisture, and area available for its growth. Plants have well-developed, branched root system. The nodes also throw out roots when come in contact with the moist soil. Leaves are simple, reniform, wavy, margin is entire having 3-5 shallow to deep lobes. The apex of the leaves may be pointed or blunt. Some genotypes have pinnatifid leaves. Bifid tendrils are borne in the axils of leaves. Commonly, it is monoecious in nature, where solitary male and female flowers are found separately in the leaf axils of the same plant. Stable andromonoecious sex form bearing hermaphrodite flower and male flower in the same plant and gynoecious lines have also been reported. Male and female flowers have large and white showy corolla with five petals. Male flowers have larger petals as compared to female flowers. In male flowers, stamens are apparently 3, two are 2-celled and one is 1-celled. Ovary is inferior and prominent, round, ovate, long or cylindrical. Ovary is tricarpillary, syncarpous and unilocular with short and stout style. There are three stigmatic lobes. Ovules are many, generally, 400-700 in number. Placentation is parietal. Fruit is a berry called pepo and a great variability is observed in fruit shapes. They may be long, cylindrical, curved-necked, oblong, round, flat-round, conical, pear-shaped, club-shaped, etc. Bottle gourd is broadly classified as long and round types. In cultivated types, long fruits vary from 40 cm to 100 cm in length at maturity, while round types may vary from 10-40 cm in diameter. The fruit weight at maturity varies from 1-10 kg. However, at edible (green) stage, the fruit weight varies from 0.5 to 2.0 kg. At edible stage, the fruits are pale green, green or dark green. Seed weight varies from 10 to 15 g per 100 seeds and individual fruit contains 400-700 seeds.

A well-drained sandy loam soil rich in organic matter with 6-7 pH is the most suitable. It is essentially a warm season crop but can be grown in regions with comperatabily low temperature. Bottle gourd require warm and humid weather for good yield. The optimum temperature is 18-20°C. Plants are very sensitive to photoperiod. The sowing season of bottle gourd is January to February for summer, and June to July for rainy season crop. In southern part of the country where temperature is mild during winter, it can be cultivated round the year. April-July is the best suitable sowing time in hills of northern India. The seed rate of bottle gourd depends upon number of seeds sown per hills. In general, the requirement of seeds/ha field is 4-6 kg. Frequent irrigation is required in spring-summer crop, while irrigation for rainy season is depends upon the rainfall. Harvesting of bottle gourd fruits is very important. Immature fruits after 10-12 days fruit setting are usually harvested. Fruits are picked at every 2-3 days. Anthracnose and Fusarium wilt are serious diseases reducing the productivity and quality, while insect red pumpkin beetle at the time of germination damaging the crop to a great extent.

Varieties

The major breeding objectives are to develop varieties/hybrids having cylindrical, small-sized (~1 kg is preferred) fruits and high yield and resistant to biotic and abiotic stresses. Most of the varieties in India have been developed through selection, hybridization, heterosis, backcross and resistant breeding. The improvement of this crop is being carried out at IIVR (Varanasi), IARI (New Delhi), IIHR, Bangalore (Bangalore), NDUA&T (Faizabad), PAU, Ludhiana (Ludhiana), GBPUA&T (Pantnagar), and CSAUA&T (Kanpur). Several varieties have been developed and described in detail as under (for photographs see plate 16).

PUSA SUMMER PROLIFIC LONG

This variety has been developed through selection from a local germplasm at IARI, New Delhi and notified by CVRC Notification no. 440 dated 21.08.1975. Its vines are vigorous in growth and bears long cylindrical fruits. It has yield potential of 250-300 q/ha.

PUSA SUMMER PROLIFIC ROUND

This variety has been developed through selection at IARI, New Delhi and notified by CVRC Notification no. 440 dated 21.08.1975. It is vigorous and prolific bearer having green round fruits 15-18 cm in girth. It has yield potential of 300-325 q/ha.

PUNJAB ROUND

This variety has been developed at PAU, Ludhiana and notified by CVRC Notification no. 13 dated 19.12.1978. Plants are vigorous and prolific bearer with round, flat, tender and shining fruits. It has been recommended for cultivation in Punjab and has yield potential of 350-375 q/ha.

KALYANPUR LONG GREEN

This variety has been developed at CSAUA&T, Kanpur and notified by CVRC Notification no. 2(E) dated 03.01.1983. Vine spreading type, fruits long, smooth, green, tender, blossom end semi round, uniform fruit size. It is suitable for rainy season and has yield potential of 275-300 q/ha.

CO-1

This variety has been developed through selection from a germplasm at TNAU, Coimbatore and notified by CVRC Notification no. 596(E) dated 13.08.1984. Its vine and leaves are yellowish green. Fruits are round at the base with a prominent bottleneck at the top, medium-sized, attractive light green and each weighing ~2 kg. It has been recommended for cultivation in Tamil Nadu and has yield potential of 360-400 q/ha in a crop duration of 135 days.

ARKA BAHAR

This variety has been developed at IIHR, Bangalore and notified by CVRC Notification no. 867(E) dated 26.11.1986. Fruits are straight, devoid of crookneck, medium-sized and each fruit weigh about 1 kg. Fruit rind is light green, flesh tender with pleasant aroma. Fruits are very good for cooking and have good keeping quality. The yield potential of this variety is 400-500 q/ha in 120 days of crop duration.

RAJENDRA CHAMATKAR

This variety has been developed at RAU, Pusa and notified by CVRC Notification no. 636(E) dated 02.09.1984. First picking starts after 68-70 days of seed sowing. Plants are prolific bearing with long and cylindrical fruits weighing 1.35 kg. It has been recommended for cultivation in Bihar and has yield potential of 175-200 q/ha.

PANT SANKAR LAUKI-1 (PBOGH-2)

This hybrid has been developed through heterosis breeding (PBOG-22 × PBOG-40) at GBPUA&T, Pantnagar and notified by CVRC Notification no. 1052(E) dated 26.10.1999. Its fruits are medium size, long and some what cylindrical (~35 cm long) and green. Vine length is ~ 5.5 m. The first picking is possible in ~60 days. It is suitable for planting in the plains as well as in hills of northern India and has yield potential of 400-425 q/ha.

PUNJAB LONG (LC-2-1)

This variety has been developed through selection from a local germplasm at PAU, Ludhiana and notified by CVRC Notification no. 1052(E) dated 26.10.1999. It is vigorous with long, tender and attractive fruits. It has been recommended for cultivation in Punjab and has yield potential of of 200-225 q/ha.

NARENDRA SANKAR LAUKI-4 (NDBGH-4)

This variety has been developed at NDUA&T, Faizabad and notified by CVRC Notification no. 597(E) dated 25.4.2006. It is early in flowering, prolific bearer with near cylindrical attractive fruits. It is suitable for cultivation throughout the year. Fruits remain soft even if they have grown out of normal size of picking. It has been recommended for cultivation throughout the country and has yield potential of 400-450 q/ha.

NARENDRA DHARIDAR (NDBG-208-1)

This variety has been developed by hybridization (NDBG-108 × NDBG-1) followed by pedigree selection at NDUA&T, Faizabad and notified by CVRC Notification no. 597(E) dated 25.4.2006. Its fruit are striped, green and bottle-shaped. It is recommended for cultivation in U.P. and has yield potential of 325-350 q/ha.

NARENDRA RASHMI (NDBG-4)

This variety has been developed through selection from a local collection at NDUA&T, Faizabad and notified by CVRC Notification no. 597(E) dated 25.4.2006. Individual plant bears ~ 8 fruits. Fruits are bottle-shaped, having shallow neck, white flash, average weight of 1.0 kg with 400-500 seeds. First harvesting starts 60 days after sowing. It is moderately resistant to powdery and downey mildew. This variety is recommended for cultivation in Punjab, U.P., Uttaranchal and Bihar and has yield potential of 275-310 q/ha.

KASHI GANGA (DVBG-1)

This variety has been developed through heterosis breeding (IC-92465 × DVBG-151) at IIVR, Varanasi and notified by CVRC

Notification no. 597(E) dated 25.4.2006. It bears light green fruits of 30 cm length, 7 cm diameter and each fruit weighing 800-900 g. It is tolerant to anthracnose and suitable for Kharif and summer season cultivation. It has been recommended for cultivation in U.P. and has yield potential of 480-550 q/ha.

KASHI BAHAR (VRH-1)

This variety has been developed through selection at IIVR, Varanasi and notified by CVRC Notification no. 597(E) dated 25.4.2006. Single plant bears 12 light green straight fruits of 30-32 cm length and each weighing 780-850g. It is suitable for rainy and summer season cultivation as well as for riverbed areas. It is tolerant to anthracnose, downey mildew and *Cercospora* leaf spot under field condition. It has been recommended for cultivation in U.P. and has yield potential of 500-550 q/ha.

NARENDRA JYOTI (NDBG-104)

This early maturing and high yielding variety has been developed at NDUA&T, Faizabad and notified by CVRC Notification no. 597(E) dated 25.4.2006. Fruits are slender and long. It has been recommended for cultivation in Punjab, U.P., Bihar and Jharkhand and has yield potential of 300-350 q/ha.

PANT LAUKI-3 (PBOG-61)

This hybrid has been developed through pureline selection from a germplasm PBOG-61 at GBPUA&T, Pantnagar and notified by CVRC Notification no. 2035(E) dated 28.11.2006. It bears ~40 cm long, cylindrical and light green fruits. First fruit harvest starts in 60 days after sowing. It is recommended for cultivation in Punjab, U.P., Bihar, Jharkhand, Rajasthan, Gujarat, Haryana and Delhi and has yield potential of 350-370 q/ha.

PANT SANKAR LAUKI-2 (PBOG-2)

This hybrid has been developed through heterosis breeding (PBOG-22 × PBOG-61) at GBPUA&T, Pantnagar and notified by CVRC Notification no. 1052(E) dated 26.10.1999. Its fruits are green, about

40 cm long and club-shaped. The first green fruit is harvested in 65 days. It is suitable for sowing in both plains and hilly areas, can be sown from March to July in plains and April to May in the hills. The seed rate is 6 kg/ha. It has been recommended for cultivation in Madhya Pradesh and Maharashtra and has yield potential of 400-425 q/ha.

AZAD SANKAR LAUKI-1

This hybrid has been developed through heterosis breeding at CSAUA&T, Kanpur and notified by CVRC Notification no. 597(E) dated 25.4.2006. Its vines are 3-3.5 m bearing long (30 cm), cylindrical and green fruits. It has yield potential of yield 580-600 q/ha.

PUSA NAVEEN

Pusa Naveen has been developed at IARI, New Delhi and identified for release through AICRP-VC in 1992. It is free from crook-necked fruits. Its fruits are cylindrical, straight and each weighing 550 g. It is suitable for cultivation in spring-summer and rainy seasons. It has been recommended for cultivation in M.P. and Maharashtra and has yield potential of 350-400 q/ha.

NDBG-132

This variety has been developed at NDUA&T, Faizabad and identified for release through AICRP-VC in 2001. It is a long fruited genotype with slender attractive fruits. It produces first fruit after 60 days of sowing. It has been recommended for cultivation throughout the country and has yield potential of 430-450 q/ha.

MUSKMELON

Cucumis melo
Family: Cucurbitaceae
Chromosome no.: 2n=24

Muskmelon, commonly known as *Kharbooj* is one of the most important dessert cucurbits of India. It is a major crop of riverbed, which covers ~80% area of total muskmelon cultivation. Muskmelon is an annual with climbing, creeping, or trailing vines, and its fruits are many-seeded pepo. It is also known as cantaloupe, a warm season vegetable that requires a little longer time than other cucurbits. Muskmelon is considered to be originated in North Africa and Asia. North West provinces of India, Kashmir, and Afghanistan, Persia, China and Southern Russia are considered to be secondary centers of diversity for muskmelon. Varieties generally having a salmon or green flesh and netted rind are usually preferred. The name describes the aroma (musk or perfume) of the ripe fruit.

The yellow and orange-fleshed melons contain carotene, which is pro-vitamin A. Cantaloupe type muskmelon is particularly high in carotene. Muskmelon is generally high in vitamin C and sugar. In melons, the aromatic compound is predominantly ester of acetic acid. Fruit contains 0.3 mg protein and 26 mg vitamin C in 100 g edible portion, and seeds contain edible oil. Immature fruits are often used as a cooked vegetable but the mature fruits are eaten raw as dessert, and occasionally used in preserves. Seeds have cooling and diuretic effects are eaten after roasting. The pulp is diuretic in nature and also used in chronic eczema. The plant extract has the blood sugar lowering principle, and roots have the emetic and purgative properties.

In India, share of muskmelon hybrids is less as compared to open-pollinated varieties. It can be grown on all types of soil but it grows best in well-drained sandy loam or loam soils rich in organic matter. Muskmelon is essentially a warm season crop mainly grown in tropical and subtropical regions. The optimum temperature to harvest good quality fruits is 27-30°C, and for proper germination the optimum day temperature should be above 25°C as the seeds do not germinate if the soil temperature is lower than 18°C. The best sowing time is 15 to 25 February in northern plains, November-March in north-eastern region, September-October in western region, and October-November in southern India. In hills of north India, sowing is done in March. In riverbeds, 2nd week of November for early crop and first week of December is the best sowing time for a late crop. The seed rate of muskmelon depends upon the sowing distance between rows and plants. The seed rate would be about 700-900 g/ha for open pollinated and 700-800 g for hybrids at recommended plant spacing keeping 3334-6000 plant per ha. Irrigation is recommended especially when the vines are growing and the fruits are developing. Excessive soil moisture during fruit ripening results poor quality fruits. Once fruits have attained good size it is best to reduce irrigation. Reduced irrigation, some times increases the sugar content of the mature fruit.

The quality of muskmelon fruit mainly depends on texture and sugar contents developed during the ripening on the vines. There are many maturity standards of ripening depending upon the cultivars and environments. Generally, fruits are harvested between 80-95 days after sowing and 30-40 days after anthesis by observing changes in outer skin color. Harvest is is generally done by hand early in the morning. Picking can be done on alternate days at the beginning of the season, and later during peak season on daily basis. Downy mildew, powdery mildew, *Fusarium* wilt and cucumber green mottle mosaic virus are main diseases causing low productivity. Among the pests, fruit fly is a major problem.

Yield is not a criterion in muskmelon, but earlyness, yield, TSS, uniformity and resistance are more considered traits while making hybrid combinations. In muskmelon, Punjab Agricultural University has been the first institution to develop Punjab Hybrid-1 followed by Pusa Rasraj.

Varieties

There are not many hybrids even from private sector. The released varieties of muskmelon have been mainly developed through selection and hybridization. The improvement of muskmelon is being progress at IARI (New Delhi), IIVR (Varanasi), PAU (Ludhiana), RAU-ARS (Durgapura), NDUA&T (Faizabad) and IIHR (Bangalore). Some of the introduced varieties have been released in India after evaluation of their adaptability to specific area/environment (for photographs see plate 17).

HARA MADHU

This variety is derived from a local collection (Kutana type) of U.P. developed at PAU, Ludhiana and notified by CVRC Notification nos. 2035(E), 440, and 1104 dated 20.02.1970, 21.08.1975, and 23.03.1978, respectively. It has long vigorous veins with sparsely lobed light green leaves. Fruits are round, tapering towards the stalk with open prominent green sutures, average fruit weight of 1 kg, not sliping from the stalk at edible maturity, thin rind, smooth and pale yellow at maturity, green flesh, thick, very juicy with 12-15% TSS. Seeds are loosely packed in the seed cavity. Post-harvest life is less with poor transportability. It has been recommended for cultivation in Punjab, U.P., Bihar, M.P. and Maharashtra and has yield potential of 130-150 q/ha.

DURGAPURA MADHU

This variety has been derived through selection from an early cultivar of Jaipur at ARS, Durgapura and notified by CVRC Notification no. 566(E) dated 21.09.1974. Fruits are oblong with big seed cavity, average fruit weight of 500-600 g; flesh is light green dry texture with 14% TSS. Durgapura Madhu has been recommended for cultivation in Punjab, U.P., Bihar, Rahasthan, Gujarat, Haryana, Delhi, M.P., Maharashtra, Karnataka, Tamil Nadu and Kerala and has yield potential of 130-150 q/ha.

LUCKNOW SAFEDA

This variety has been developed at PAU, Ludhiana and notified by CVRC Notification no. 440 dated 21.08.1975. It is an early variety with medium vine growth. Fruits are small, roundish, flat, yellow, smooth and each weighing ~300 g. Flesh is creamy white, less juicy and sweet with good keeping quality. It has yield potential of 110-130 q/ha.

PUSA SHARBATI

This variety has been developed from a cross Kutana × American Cantaloupe (PMR-6) at IARI, New Delhi and notified by CVRC Notification no. 1004(E) dated 23.03.1978. Fruits are round with nettled skin, salmon orange flesh, firm, thick, 11-12% T.S.S. with small seed cavity. It has good keeping quality and suitable for riverbed cultivation. It has been recommended for cultivation in riverbeds and upland conditions of northern Gangatic plains and has yield potential of 150-170 q/ha.

PUNJAB SUNEHRI

This variety has been developed through hybridization (Hara Madhu × Edisto) followed by pedigree selection at PAU, Ludhiana and notified by CVRC Notification no. 13 dated 19.12.1978. Its vines are medium-long with lobed leaves, dark green with short petioles. Fruits are round to elliptical and devoid of distinct sutures at maturity. Avarage fruit weight is 1 kg. It slips easily from the stalk at edible maturity. Its rind is thick, netted and dull yellow at maturity. Flesh is thick, orange, fairly juicy and sweet (11% TSS). It has a long post-harvest life and has excellent keeping quality. It is tolerant to powdery and downy mildew and has yield potential of 160-180 q/ha.

GUJARAT MUSKMELON-1

This variety has been developed through selection from a local collection from Sabarkantha district in Gujarat at GAU, Anand and notified by CVRC Notification no. 596(E) dated 30.08.1984. It has 5-6 fruits/plant, which are small, tasty, lemon yellow, smooth and without suture. Flesh is light green with 2.1 cm thickness. The average weight of fruit is ~1.0 kg with a diameter of 10 cm. It has yield potential of 140-170 q/ha.

GUJARAT MUSKMELON-2

This variety is selection from a local germplasm collected from Nagpur area of Maharashtra, developed at GAU, Anand and notified by CVRC Notification no. 596(E) dated 30.08.1984. Flowering starts at 32 days after sowing. It has 4-5 fruits/plant, which are medium in size (diameter 11.5 cm) and each weighing 1.3 kg. Skin is orange green with slight netting. Flesh is capucine orange with 2-6 cm thickness. It has yield potential of 120-150 q/ha.

ARKA RAJHANS

This variety has been derived through single plant selection from a local collection (IIHR-107) of Rajasthan, developed at IIHR, Bangalore and notified by CVRC Notification no. 867(E) dated 26.11.1986. Plants have densely hirsute stem, light green foliage with shallow lobes and bear 2-3 fruits. Fruits are round to slightly oval, medium-large, creamy white skin, fruit weight is 1.25-2.0 kg, netted skin with shallow vein tracks at full slip stage, thick, firm and white flesh with 11-14% of TSS, crisp texture, mild flavour, good transport and keeping quality. It is resistant to powdery mildew. It has been recommended for cultivation in Karnataka, Tamil Nadu and Kerala and has yield potential of 285-300 q/ha in 85-90 days of crop duration.

ARKA JEET

This variety has been derived through single plant selection from a local collection (Bati Strain; IIHR-103) of Lucknow, developed at IIHR, Bangalore and notified by CVRC Notification no. 867(E) dated 26.11.1986. It has dark green lobed leaves and angular stem. Fruits each of 300-500 g are small and round with smooth orange flat ends, dark green skin, thin golden orange at full slip stage. Flesh is white with medium soft texture, excellent flavour, TSS 15-17%, juicy and rich in vitamin C. It has been recommended for cultivation in Karnataka, Tamil Nadu and Kerala and has yield potential of 150-175 q/ha in 90 days of crop duration.

PUNJAB HYBRID-1

This hybrid has been developed through heterosis breeding (ms-1 × Hara Madhu) at PAU, Ludhiana and notified by CVRC Notification no. 834(E) dated 18.09.1987. It has vigorous, medium long vines with lobed leaves. Fruits are globular with distinct sutures. Avarage fruit weight is ~800 g and slips from the stalk at edible maturity stage. Rind is quite thick, netted and creamy yellow. Flesh is thick, orange and juicy. It is sweet with 12% TSS. It is moderately resistant to powdery mildew and fruit fly. It has been recommended for cultivation in Punjab, Bihar, U.P., Karnatka, Tamil Nadu, and Kerala and has yield potential of 160-180 q/ha.

PUSA RASRAJ

This hybrid has been developed at IARI, New Delhi, and notified by CVRC Notification no. 617(E) dated 17.08.1993. Fruits oval, skin smooth without netting, weighing 800 g to 1 kg, sweet (TSS 11%), maturity 75-80 days. This hybrid has yield potential of 225-240 q/ha.

PUNJAB RASILA (MR-12)

This variety has been developed at PAU, Ludhiana and notified by CVRC Notification no. 115(E) dated 10.02.1996. It has long vines with fruit setting close to the base of vine. Its fruits are nearly round, medium netting with light yellow rind and green sutures. Flesh is green, thick, juicy, 10% TSS and flavoursome. Fruit develops full-slip stage at maturity. Average fruit weight is 600 g. It takes 73 days from transplanting to first picking (10 days earlier than Hara Madhu). It has yield potential of 160-175 q/ha.

MH-10

This hybrid has been developed through heterosis breeding (WI-998 × Punjab Sunehri) at PAU, Ludhiana, and notified by CVRC Notification no. 115(E) dated 10.02.1996. It is early maturing and fruit setting takes place close to the base of vine, fruits are round/oval, rind netted, non-sutured, thick, brown on maturity with average weight of 915 g, flesh is thick, orange, medium juicy, 910% TSS, fruit developed full-slip stage at maturity, good shelf-life and suitable for distant transportation. It has yield potential of 240-250 q/ha.

RM-43

This variety has been developed through a selection from local collected germplasm at RAU-ARS, Durgapura and notified by CVRC notification no. 98(E) dated 08.02.1997. It bears oblong pale green fruits with dark green 10-12 sutures, each weighing 400-600 g, hard rind, small seed-cavity, green flesh, and 10-12% TSS. Fruits have good keeping quality and transportability. It has yield potential of 200-220 q/ha.

MHY-3

This variety has been developed through selection from a cross (Durgapura Madhu × Pusa Madhuras) at RAU, ARS, Durgapura, and notified by CVRC Notification no. 1052(E) dated 26.10.1999. It vines are vigorous and spreading about 2.5-3.0 m long, and fruits are round flat, smooth surfaced with yellow rind and light green stripes. Flesh is light green, texture soft, and 13-16% TSS. Average fruit weight is 700-800 g. It has been recommended for cultivation in Rajsthan and has yield potential of 180-200 q/ha.

RM-50

This variety has been developed through hybridization (Durgapura Madhu × Sel-1) followed by pedigree selection at RAU, ARS, Durgapura, and notified by CVRC Notification no. 642(E) dated 31.05.2004. Its vines are well developed up to 2-2.5 m. Fruit shape ovate, pale green in colour having ten green sutures, and 14-16% TSS with green flesh. Average fruit weight is 500 g. It has been recommended for cultivation in Rajsthan and has yield potential of 175-200 q/ha.

MHY-5

This variety has been developed through hybridization (Durgapura Madhu × Hara Madhu) followed by pedigree selection at RAU, ARS, Durgapura, and notified by CVRC Notification no. 642(E) dated 31.05.2004. Its vines are vigorous and moderately spreading about 2-2.5 m long. Fruits with smooth rind and medium seed cavity are round flat, tapering at stem end and green yellow in colour. The flesh is light green, texture soft, and TSS 13-16%. Average fruit weight is 700-800 g. It has been recommended for cultivation in Madhya Pradesh and Maharashtra and has yield potential of 180-200 q/ha.

GUJARAT MUSKMELON-3

This variety has been developed from a local collection at Anand Agricultural University, Anand and notified by CVRC Notification no. 597(E) dated 25.4.2006. Its fruits are medium size, round to slightly oval, yellowish brown spots on mature fruits with green strips. Flesh is green, juicy, very sweet, 12% TSS, average fruit weight 800 g. It has been recommended for summer cultivation in irrigated areas of middle Gujarat and has yield potential of 110-125 q/ha.

KASHI MADHU (IVMM-3)

This variety has been developed from a local collection of U.P. at IIVR, Varanasi and notified by CVRC Notification no. 2035(E) dated 28.11.2006. It has medium vine and sparsely lobed and dark green leaves. Fruits are round, with open prominent green sutures, weight 650-725 g, half-slip in nature, thin rind, smooth and pale yellow at maturity, flesh salomon orange, thick, very juicy with 13-14% TSS and seeds are loosely packed in the seed cavity. Post-harvest life is long with good transportability. It is tolerant to powdery and downy mildew. It is recommeneded for cultivation in Punjab, U.P., Bihar, Jharkhand and Uttaranchal and has yield potential of 250-270 q/ha.

NARENDRA MSUKMELON-1 (NDM-2)

This variety has been developed at NDUA&T, Faizabad and notified by CVRC Notification no. 597(E) dated 25.04.2006. Its fruits are round, large-sized with orange colour skin, green stripped at maturity. Flesh is dark orange, thick with 9-11% TSS. Single fruit weight 700-900 g. It is recommeneded for cultivation in Uttar Pradesh and has yield potential of 125-150 q/ha.

PUSA MADHURAS (SEL-45)

This variety has been derived through selection from a germplasm at IARI, New Delhi and identified for release through AICRP-VC in 1975. Fruits are roundish-flat, each weighing 1 kg, sparsely netted with pale green strips, flesh is salmon-orange, juicy with 12-14% TSS. It has poor keeping quality. First picking starts 90-95 days after sowing. It has been recommended for cultivation in Punjab, U.P., Bihar, Rahasthan, Gujarat, Haryana, Delhi, M.P., Maharashtra, Karnataka, Tamil Nadu and Kerala and has yield potential of 150-170 q/ha.

NARENDRA MSUKMELON-2 (NDM-15)

This medium maturity variety has been developed at NDUA&T, Faizabad and identified for releae through AICRP-VC in 2003. Fruits are oval-round with smooth skin, orange coloured with gree striped at maturity, flesh is thick, red-orange, 10-13% TSS. Single fruit weight 800-1000 g and first fruit can be harvested in 70-75 days. It is recommeneded for cultivation in Punjab, U.P., Bihar and Jharkhand and has yield potential of 150-200 q/ha.

BITTER GOURD

Momordica charantia
Family: Cucurbitaceae
Chromosome no.: 2n=22

Bitter gourd (syn. bitter melon, balsam pear, bitter cucumber) known as karela in Hindi, *Karla* in Marathi, *Kakara kaya* in Telugu, *Pavakai* in Tamil, *Agala kayi* in Kannada and *Pavakka* in Malayalam. It is an important cucurbit grown mainly in tropics and subtropics. It is widely grown in India, Indonesia, Malaysia, Singapore, China, Japan, Southeast Asia, tropical Africa and South America. In South India, this crop is grown on commercial scale and exported to Gulf and European countries. Some of the leading states cultivating bitter gourd in India are Uttar Pradesh, Orissa, Maharashtra, Andhra Pradesh, Tamil Nadu and Kerala. The origin of bitter gourd is not known, however, it is considered to be a native of tropical Asia, particularly eastern India and southern China.

Tender fruits of bitter gourd are used as a cooked vegetable or pickled. Bitter taste of fruit is due to alkaloid momordicin. Tender shoots and leaves are also consumed as green vegetable after parboiling to leach out the bitter alkaloid. Seeds of the ripe fruit are used as a condiment in India. Steeping the peeled fruit in salt water before cooking can reduce its bitterness. Bitter gourd is not advocated for use to pregnant or nursing women. The fruits, leaves and roots have several beneficial medicinal uses, especially, in treatment of diabetes. It has property of analgesic, abortifacient, immunosuppressive and anti-tumor properties. Fruits are rich in iron, calcium, phosphorus, vitamin A and C.

A well-drained loamy soil rich in organic matter having 6.6-7.0 pH is suitable for cultivation of bitter gourd. It is essentially a warm season crop but can be grown in regions with comperatabily low temperature. Plants are verey sensitive to frost. The optimum temperature for vegetative and reproductive growth is 18-35°C. In northern plains seed can be sown from January to February for summer crop, while June to July is suitable for rainy season crop. In southern part of the country where temperature is mild during winter, it can be cultivated round the year. April-July is suitable for seed sowing in hills of northern India. The seed requirement for 1 hectare area is 4-5 kg. In summer, crop should be irrigated at weekly interval. Immature tender fruits should be harvested. Downy mildew and tospo virus (caused by Gemini virus) are serious diseases reducing the productivity and quality. Red pumpkin beetle at the time of germination and fruit fly (at early fruit development stage) damaging the crop to a great extent.

Varieties

Most of the varieties of bitter gourd have been developed through selection from germplasm, hybridization followed by pedigree selection or by heterosis breeding. Induced mutation has also been used for development of varieties. Genetic improvement of bitter gourd is being carried out at IIHR (Bangalore), IIVR (Varanasi), IARI (New Delhi), KAU (Vellanikkara), MPKV (Rahuri) and CSAUA&T (Kanpur). Several varieties have been developed and described in detail as under (for photographs see plate 18).

COIMBATORE LONG

This variety has been developed at TNAU, Coimbatore and notified by CVRC Notification no. 440 dated 21.08.1975. Fruits are long, tender and white. It is suitable only for rainy season cultivation and has yield potential of 200-220 q/ha.

CO-1

This variety has been developed through a selection from a local genotype collected from Thudiyalur at TNAU, Coimbatore and notified by CVRC Notification no. 2103 dated 12.08.1980. Vines are medium long (130 cm) with 5 to 7 branches. First flowering starts in 45-50 days and marketable fruit is harvested in 55-60 days after sowing. A total of 6-8 picking can be done. Fruits are dark green, medium long (25-30 cm) and thick (6-8 cm) with characteristic warts. Each vine produces 20-22 fruits of an average weight 100-120 g. This variety has yield potential of 140-150 q/ha.

KALYANPUR BARAMASI

This variety has been developed at CSAUA&T, Kalyanpur and notified by CVRC Notification no. 2(E) dated 03.01.1983. Its vines are vigorous with long (30-50 cm) light green, thin and tapering fruits. It is tolerant to fruitfly and mosaic virus. It is recommended for kharif season cultivation in U.P. and has yield potential of 200-220 q/ha in the crop duration of 120 days.

MDU-1

This variety has been developed through mutation breeding (gamma irradiation of local cultivar MC-103) at TNAU, Coimbatore and notified by CVRC Notification no. 540(E) dated 24.07.1985. Flowering starts at 60 days after sowing with a sex ratio of 1:20 of female and male flowers. Fruits are long with an average length of 40.34 cm and a girth of 17.54 cm having average fruit weight of 410 g. Each vine bears on an average of 16-18 fruits. It has been recommended for cultivation in Tamil Nadu and has yield potential of 300-325 q/ha.

ARKA HARIT

This variety has been developed at IIHR, Bangalore through selection from local genotype collected from Rajasthan and notified by CVRC Notification no. 867(E) dated 26.11.1986. Fruits are spindle shaped, attractive, glossy green with smooth regular ribs and thick flesh. It has been recommended for cultivation in Karnataka and has yield potential of 125-135 q/ha in the crop duration of 120 days.

PUNJAB-14

This variety has been developed through selection from local materials at PAU, Ludhiana and notified by CVRC Notification no. 165(E) dated 06.03.1980. Its vines are comparatively short having large, oblong and green fruits. It is suitable for spring summer and rainy season cultivation. It has been recommended for cultivation in Punjab and has yield potential of 120-130 q/ha.

KALYANPUR SONA

This variety has been developed at CSAUA&T, Kalyanpur and notified by CVRC Notification no. 115 dated 10.02.1996. Its vines are medium in growth having medium sized and green fruits. This variety has been recommended for cultivation in Uttar Pradesh and has yield potential of 120-130 q/ha.

PREETHI (MC-84)

This variety has been developed at KAU, Vellanikkara and notified by CVRC Notification no. 843(E) dated 21.09.1998. It has light green vine of ~7 m length. Fruits are white, spiny, medium in size with 310 g. It has been recommended for cultivation in M.P., Maharashtra, Karnataka, Tamil Nadu and Kerala and has yield potential of 140-150 q/ha in 150 days of crop duration.

PHULE PRIYANKA (RHRBGH-1)

This hybrid has been developed through heterosis breeding (RHR-5 × RHR-4) at MPKV, Rahurui and notified by CVRC Notification no. 3485(E) dated 19.04.2001. Fruits are medium long, dark green, and highly prickly. It is tolerant to downy mildew and has yield potential of 200-250 q/ha in 180 days of crop duration.

PHULE GREEN GOLD (RHRBG-5)

This variety has been developed through hybridization (Green Long × Delhi Local) followed by pedigree selection at MPKV, Rahuri and notified by CVRC Notification no. 348(E) dated 19.04.2001. Fruits are dark-green, 25-30 cm long with raised tubercles. It is tolerant to downy mildew. It has been recommended for cultivation in Maharashtra and has yield potential of 220-235 q/ha in crop duration of 150-180 days.

KONKAN TARA (DPLBG-2)

This variety has been developed at KKV, Konkan and notified by CVRC Notification no. 115 dated 10.02.1996. Its fruits are green, medium long (15-16 cm) and spindle-shaped with raised tubercles and good keeping quality with a shelf-life of 7-8 days at ambient temperature. It has been recommended for cultivation in Konkan region of Maharashtra and has yield potential of 230-250 q/ha.

PANT KARELA-2 (PBIG-1)

This variety has been developed through selection from a germplasm PBIG-1 at GBPUA&T, Pantnagar and notified by CVRC Notification no. 2035(E) dated 28.11.2006. Its fruits are thin, 25 cm long, dark green with tapering ends. First marketable fruit picking starts at 50 days after seed sowing. It has been recommended for cultivation in Uttar Pradesh, Uttarakhand, Bihar and Chattisgarh and has yield potential of 150-160 q/ha.

PRIYA (VK-1)

The variety Priya has been developed at KAU, Vellanikkara and identified for release through AICRP-VC in 1992. Its fruits are long (40 cm), spiny, green with white tinge at the style end. It can be cultivated during January-August and Septemder-December. It has been recommended for cultivation in West Bengal, Assam, M.P., Maharashtra, Karnatka, Tamil Nadu and Kerala and has yield potential of 300-325 q/ha.

PUSA HYBRID-2

This hybrid has been developed by heterosis breeding utilizing inbred Pusa Vishesh and Pusa Do Mausami at IARI, New Delhi and identified for release through AICRP-VC in 2002. Its fruits are medium long and thick, glossy-green, suitable for pickling and dehydration. First picking starts at 55-60 days after seed sowing. It has been recommended for sowing in spring-summer season in Punjab, U.P., Bihar, Jharkhand, Chhatishgarh, Orissa and A.P., Rajasthan, Gujarat, Haryana and Delhi and has yield potential of 190-200 q/ha.

CUCUMBER

Cucumis sativus
Family: Cucurbitaceae
Chromosome no.: 2n=14

Cucumber is a warm season annual that prefers hot humid weather. It is mainly grown for its immature fruits and primarily used as salad. In addition, this is pickled and its young leaves and stems are cooked in Southern Asia. Seeds are used in a variety of ways for human consumption. Cucumber is low in calories (113-148 K Cal per fruit) and also low in mineral and vitamin value. Excellent flavor and texture of fruits make cucumbers most priced salad vegetable.

Cucumber cultivars are usually classified according to their intended use. Fruits for fresh market slicing are preferably long, smooth, straight, and thick skinned, with a uniform dark green colour. Pickling fruits are light green in colour, small-sized, thin-skinned and characterized by a warty surface. Fresh market cultivars have fewer spines than processing types. Regardless of how they are to be used, cultivar differences in earliness and disease resistance are also important considerations for their selection.

On the basis of occurrence of pistil or stamens, five types of sex forms have been reported in cucumber, *viz*; gynoecious, hermaphrodite, monoecious, gynomonoecious and androecious. Out of these sex forms, only gynoecious and monoecious types are utilized in breeding programmes. Generally, gynoecious sex form is available in temperate countries and it is very difficult to maintain gynoecy beyond 30°C in tropical countries like India. This character now has been incorporated in tropical background and at IARI Regional Station, Katrain, a stable gynoecious line has been developed. Anthesis of flowers take place between 5-30-7.00 A.M. and anther dehiscence occurs early in the morning before anthesis (4:30-5:00 A.M.). Duration of pollen fertility is up to 14 hours after anthesis and stigma remains receptive 12 hours before and 10 hours after anthesis.

A well-drained light textured soil with proper drainage is suitable for cucumber cultivation. Among the cucurbits, cucumber is considered most sensitive to frost. The optimum temperature for proper growth and development is 18-30°C. The sowing time of cucumber in northern plains is January to February for summer crop and June to July for rainy season crops. In the southern parts of the country, it can be cultivated round the year due to mild temperature during winter season. April to June is the best sowing time in hilly regions of northern India. The seed requirement for 1 hectare area is 3-4 kg. Irrigation at shorter interval is required during summer season. Fruits are harvested at right edible maturity stage. The marketable maturity of fruits depends upon the varieties and growing season. Fruits become ready for harvesting after 7-10 days of fruit setting. Downy mildew, powdery mildew and cucumber mosaic virus are main diseases of cucumber. Among the pests, leaf minor and mites are becoming the problem under protected conditions.

In genetic improvement programmes, uniformity in fruit quality is very much desired by consumers. Uniformity in size, shape and colour (of the outer skin) of the cucumber fruits can easily be obtained in hybrids, under controlled conditions. Breeding objectives are often a regional goal because of the preference of fruit colour, shape and size. In the light of advances in protected cultivation technologies, there has been demand of cucumber hybrids, specially suited for cultivation under protected conditions. In addition, high yielding early variety/ hybrid with light green to green skin, smooth fruits, lack of bitterness in fruit near the stem end at edible stage, cylindrical fruit shape without crooked neck and multiple disease resistance are prime goal.

Varieties

The genetic improvement of cucumber in India is being carried-out at IIVR (Varanasi), IARI (New Delhi) and its regional station, Katrain, HARP (Ranchi), YSPUH&F (Faizabad), PAU (Ludhiana), GBPUA&T (Pantnagar), and CSAUA&T (Kanpur). Several varieties of different maturity groups have been developed and described in detail as under (for photographs see plate 19).

JAPANESE LONG GREEN

This variety has been developed at IARI, Regional Station, Katrain and notified by CVRC Notification no. 440 dated 21.08.1975. It is early, prolific bearing and first picking starts at 45 days after sowing. Fruits are yellowish green, 30-40 cm long, whitish green with light green and crisp flesh. It essentially requires staking for straight fruits. It is suitable for cultivation in hilly and mild climates and has yield potential of 150-275 q/ha.

SOLAN GREEN

This variety has been developed at YSPUH&F, Solan and notified by CVRC Notification no. 440 dated 21.08.1975. Its fruits are light green and 11-15 cm long. It has yield potential of 180-200 q/ha.

KALYANPUR GREEN

This variety has been developed at CSAUA&T, Kanpur and notified by CVRC Notification no. 2(E) dated 03.01.1983. It is spreading type and early in fruiting. Fruits are thick, green but brown at maturity. It is suitable for cultivation in rainy season and has yield potential of 100-125 q/ha.

SHEETAL

This early maturing and high yielding variety has been developed at KKV, Dapoli and notified by CVRC Notification no. 915(E) dated 06.11.1989. Its vine measures upto 2.7 m with upto 10 branches/vine. Its fruits are green, medium in length (~25 cm) and 19.6 cm girth and each weighing 300-350 g. It bears 24 fruits/plant in a crop duration of 95 days. It can be raised from June to September, while the summer crop may be sown in January-February and has yield potential of 380-400 q/ha in rainy season and 200-230 q/ha in summer season.

PHULE SUBHANGI (SEL-75-1-10)

This variety has been developed through pedigree selection at MPKV, Rahuri and notified by CVRC Notification no. 348(E) dated 19.04.2001. Its fruits are smooth and attractive green fruits with white strips at apical end. It is tolerant to pod borer, anthracnose, leaf spot and leaf miner under field conditions and has yield potential of 300-325 q/ha in 90 days of crop duration.

GUJARAT CUCUMBER-1

This variety has been developed from a local collection at GAU, Anand and notified by CVRC Notification no. 597(E) dated 25.4.2006. Its fruits are longer, tender and attractive with light green strips, soft and non-hairy with longer shelf-life. Gujrat Cucumber-1 has been recommeneded for cultivation in Gujarat and has yield potential of 240-260 q/ha.

SWARNA POORNA

This variety has been developed at HARP, Ranchi and notified by CVRC Notification no. 642(E) dated 31.05.2004. Its fruits are cylindrical, long, medium-sized (~300g), light green without placental hollowness. It is tolerant to powdery mildew. Fruit picking starts at 55-60 days after sowing. It has yield potential of 310-335 q/ha in 100-120 days of crop duration.

SWARNA AGETI

This variety has been developed through mutation breeding at HARP, Ranchi and notified by CVRC Notification no. 2035(E) dated 28.11.2006. Its fruits are cylindrical, long, medium size (~200g), green without placental hollowness. It is tolerant to powdery mildew. First picking starts at 45-50 days after sowing. It has yield potential of 300-325 q/ha in 100-120 days of crop duration.

PANT KHIRA-1 (PCUC-28)

This variety has been developed through selection from an indigenous germplasm at GBPUA&T, Pantnagar and notified by CVRC Notification no. 2035(E) dated 28.11.2006. Its fruits are long (~20 cm), cylinderical with light white stripes. First picking starts at 50-60 days after sowing. Seed rate is 4 kg/ha and the optimum planting distance is 150 cm × 30 cm. It is suitable for planting in both plains (March-July) and hills (April-May) and has yield potential of 130-150 q/ha.

SWARNA SHEETAL

This variety has been developed at HARP, Ranchi and notified by CVRC Notification no. 2035(E) dated 28.11.2006. Its fruits are cylindrical, long, medium size (~250 g) with whitish green without placental hollowness. First picking starts at 60-65 days after sowing and the variety has yield potential of 400-450 q/ha in 100-120 days of crop duration.

PANT SANKAR KHIRA-1 (PCUCH-1)

This hybrid has been developed through heterosis breeding (PCUC-28 x PCUC-8) at GBPUA&T, Pantnagar and identified for release through AICRP-VC in 2001. Fruits are long (~20 cm), cylindrical and green with white stripes. Its vine length is ~120 cm. It is suitable for planting in the plain as well as hills and it takes 50 days to first picking. It is recommeneded for cultivation throughout the country. It has yield potential of 180-200 q/ha.

PCUCH-3

This hybrid has been developed through heterosis breeding (PCUC-15 x PCUC-25) at GBPUA&T, Pantnagar and identified for release through AICRP-VC in 2005. Fruits are cylindrical, green and medium long (about 20 cm) with white spots. It has long fruiting period and first fruit can be harvested in 50 days. It is moderately resistance to downy mildew and powdery mildew. It is recommeneded for cultivation in J&K, H.P., Uttaranchal, Punjab, U.P., Bihar and Jharkhand. This hybrid has the yield potential of 250-270 q/ha.

PUMPKIN

Cucurbita moschata
Family: Cucurbitaceae
Chromosome no.: 2n=40

Pumpkin is one of the most popular cucurbit vegetables grown commercially all over India. Pumpkin is also known as *kashiphal* or *sitaphal* or *lal kaddu*. It is extensively grown in different parts of the country for its immature and mature fruits. As a vegetable, pumpkin is of great significance due to high carotene content and good keeping quality. The genus *Cucurbita* is a native of Central or South America. Archaeological evidences suggest that *C. moschata* is widely distributed in America, Mexico and Peru, being the primary centers of origin. The pumpkins appeared in Europe in mid-16th century. It was introduced to India from South America by foreign navigators and emissaries. Its mature ripe fruits have pale reddish brown skin with ribbed or smooth surface. Its immature fruits of greenish or greenish yellow skin are also harvested and marketed. Fruits of pumpkin are mostly oblong, round, sometimes oval and weight varies from 2-20 kg. The flesh colour ranges from pale yellow to crimson and flesh thickness from 3.5-5.5 cm.

There are five cultivated species of *Cucurbita*, *viz*., *C. moschata* (Pumpkin), *C. pepo* (Summer squash), *C. maxima* (winter squash), *C. mixta* (syn. *C. angyrosperma*; winter squash) and *C. ficifolia* (Malabar gourd or fig leaf gourd). Among these, *C. moschata*, *C. pepo* and *C. maxima* are of great economic importance. Commonly all *Cucurbita* species are called as pumpkin; however their immature fruits are known as squash.

Pumpkins are used as fresh vegetable, processed or stock feed. Both, immature and mature fruits are consumed raw, cooked, stewed, boiled or baked. Fruit are also used in the cure of stomach catarrh, hyperacidity and even stomach ulcers. Pumpkin raw pulp has purgative effect. Pumpkin is recommended incase of obesity and deterioration in the function of gall bladder. Pumpkin juice and raw/ boiled fruits have diuretic and cholagogue properties. Its juice or decoction with honey is also advisable for people suffering from insomnia or from uneasy and troubled sleep. Pumpkin seeds contain about 52% of high quality edible oil.

Pumpkin can be grown in well-drained loam soil. Sandy loam soil rich in organic mathes is preferred for growing early crop. Soil should be rich in organic matter. Pumpkin is moderately tolerant to soil acidity (up to pH 5.5), and grow well in a pH range of 5.5-7.0. Being a warm season vegetable and can not tolerate frost. Temperatures range of 25-30°C in day and 20-25°C in night, are favorable for proper growth, flowering and fruiting. Plants are usually killed by one hour or more of frost (temperature below 0°C). Therefore, it should be sown or planted when soil temperatures are high enough for good germination.

Seeds are sown during January-March for summer season, whereas for rainy season crop, they are sown in June-July. In South and Central India, where winter is mild, the crop is grown almost throughout the year. In the Western Gangetic plains, early sowing is done in riverbed from November to February. Pumpkin seeds are sown at 3.0-3.5 m row-to-row and 1.0 m plant-to-plant spacing. Around 3000 plants can be accommodated in one hectare with seed rate of 5-7 kg. Seeds germinate in 7-8 days and after germination, two healthy seedlings/pit are allowed to grow. Light and frequent irrigation is very essential in spring-summer crop, while in rainy season crop, irrigation may not be essential at all, if rain is well distributed during June to September. Water stress during these phases may reduce plant growth, flowering and fruiting. Moisture deficiency during flowering and fruiting may cause wilting and drying of apical portion of the developing fruits. Red pumpkin beetle and fruit fly are the major insects and downy mildew, powdery mildew and pumpkin mosaic virus are the main diseases damaging the crops.

Varieties

Most of the varieties of pumpkin have been developed through selection or hybridization followed by pedigree selection. Heterosis breeding has also been utilized to develop hybrids of superior quality and yield. In India, the genetic improvement of pumpkin is being carried out at IARI (New Delhi), IIVR (Varanasi), NDUA&T (Faizabad), KAU (Vellanikkara), GBPUA&T (Pantnagar), IIHR, Bangalore (Bangalore), TNAU (Coimbatore) and CSAUA&T (Kanpur) and several varieties have been developed and described in detail as under (for photographs see plate 20).

CO-2

This variety has been developed through selection from a local collection at TNAU, Coimbatore and notified by CVRC Notification no. 13 dated 19.12.1978. Its vines are moderately vigorous and less-spreading, distance for high density planting 2.2 × 1.9 m. Fruits are small, slightly ridged with bright orange flesh, 10.8% TSS, 3.4% total sugar, 1.9% reducing sugar, 9.2% starch, 10.0 mg/100g ascorbic acid, and 0.14% acidity. First picking starts at 95-100 days after sowing. It has yield potential of 450-475 q/ha in 135 days of crop duration.

AMBILI (CM-14)

This variety has been developed through selection from a local cultivar at KAU, Vellanikkara and notified by CVRC Notification no. 1135(E) dated 01.12.1988. It is spreading type with fruits flat-round, medium (~6 kg weight), green with shallow furrows on surface. It has been recommended for cultivation in states of Kerala, Tamil Nadu, Karnataka and has yield potential of 340-360 q/ha in 95-110 days of crop duration.

PUSA BISWAS

This variety has been developed through selection from a local collection (Line SM-107) at IARI, New Delhi and notified by CVRC Notification no. 386(E) dated 15.05.1990. It has vigorous vegetative growth, dark green leaves with white spots including veins. Fruits are brown, spherical each weighing ~5 kg with thick and golden yellow flesh. It has been recommended for cultivation in J&K, H.P., Uttaranchal, Punjab, U.P., Bihar, Chhattisgarh, Orissa, A.P., M.P. and Maharashtra and has yield potential of 400-425 q/ha in 120 days of crop duration.

ARKA CHANDAN (IIHR-105-1-1)

This variety has been developed through selection from a local collection (IIHR-105) at IIHR, Bangalore and notified by CVRC Notification no. 386(E) dated 15.05.1990. It has vigorous vines, hairy stem, green broad leaves without any pattern. Fruits are round with flat blossom end and each of 2-3 kg weight, green with white patches on rinds (turn light brown at maturity), thick flesh, and bright orange firm with pleasant aroma, carotene rich and sweet with 8-10 % TSS. It has been recommended for cultivation in Karnataka, Tamil Nadu and Kerala and has yield potential of 325-350 q/ha in 120 days of crop duration.

KASHI HARIT (IVPK-226)

This variety has been developed through hybridization (NDPK-24 x PKM) followed by pedigree selection at IIVR, Varanasi and notified by CVRC Notification no. 597(E) dated 25.04.2006. It has short vines and dark green leaves with white spots. Fruits are green, spherical, weight 2.5-3.0 kg at green stage. It has been recommended for cultivation in U.P., Bihar, Chhattisgarh and Uttaranchal and has yield potential of 300-350 q/ha in 65 days of crop duration.

SOORAJ (CM-350)

This variety has been developed at KAU, Vellanikkara and notified by CVRC Notification no. 448(E) dated 21.04.2003. It produces medium-size globular fruits with orange coloured thick flesh. Average fruit weight is ~3 kg. It is recommeneded for cultivation in West Bengal, Assam, Karnataka, Tamil Nadu and Kerala and has yield potential of 350-400 q/ha.

NARENDRA AGRIM (NDPK-24)

This variety has been developed at NDUA&T, Faizabad and notified by CVRC Notification no. 597(E) dated 25.04.2006. It has short vines, very early maturing, small fruited and highly homogeneous. Fruits round, dark green. It bears female flowers four days earlier than male flowers and first picking could be taken within 55 days. Green fruits weigh 1.5-2.0 kg and mature fruits are of 3-4 kg weight. It is recommeneded for cultivation in Punjab, U.P., Bihar, Jharkhand, Rajasthan, Gujarat, Haryana and Delhi and has yield potential of 300-350 q/ha.

AZAD PUMPKIN-1

This variety has been developed at CSAUA&T, Kanpur and notified by CVRC Notification no. 597(E) dated 25.4.2006. It is spreading type with green hairy leaves, early fruiting, close fruit bearing, green fruit with yellowish stripes, thick flesh, less seeds and spherical flat. It has been recommended for cultivation in U.P. and has yield potential of 425-450 q/ha.

NARENDRA AMRIT (NDPK-130)

This variety has been developed at NDUA&T, Faizabad and notified by CVRC Notification no. 597(E) dated 25.04.2006. First picking starts within 60 days in summer and bears female flowering earlier than male flowers. Fruits are round, pale-green-mottled, turning almond colored at maturity. Single green fruit is of ~3 kg, while mature fruits are of 6 kg weight. Mature fruits have papery skin, thick flesh of about 7 cm with small seed cavity. Both green and mature fruits can be used as *salad,* while mature fruits can be used for nutritious fruit juice and *Halwa/Burfi* preparations. Mature fruits have long storage life and can be kept safely for more than four months at ambient temperature. It has yield potential of 350 to 400 q/ha.

SARAS

This variety has been developed at KAU, Vellanikkara and notified by CVRC Notification no. 597(E) dated 25.04.2006. It has elongated, orange fleshed and medium size fruits each fruit weighing ~2.7 kg. It has yield potential of 390-410 q/ha.

NARENDRA ABHOOSHAN (NARENDRA PUMPKIN HYBRID-1)

This hybrid has been developed at NDUA&T, Faizabad and notified by CVRC Notification no. 597(E) dated 25.04.2006. It is an early prolific bearing near round fruits with dark green stripes, turning deep orange at maturity. The variety is suitable for mid January to mid March sowing. It has been recommended for cultivation in U.P. and has yield potential of 650 to 700 q/ha.

SQUASH

Cucurbita pepo, C. maxima
Family: Cucurbitaceae
Chromosome no.: 2n=40

Summer squash (syn. vegetable marrow) is called chappan kaddu. Plants are fast growing, early in fruiting and relatively tolerant to low temperature. The variations in fruit shape and colour make it attractive to the consumers. There is a confusion in the name of *Cucurbita* spp.; the species generally grown for immature fruits are referred to as summer squash, whereas those produced for mature fruits are called winter squash or pumpkin. Fruits are generally consumed immature as vegetable, in some parts of the country (West Bengal, Orissa etc.). Growing tips and flowers are also consumed. It was reported in north eastern Mexico in early 7000 BC. *C. pepo* was originated in north eastern Mexico of America. It is now widely distributed throughout world. There is no evidence of its cultivation in India.

It is annual with deeply lobed leaves. Petiole is covered with prickly hairs and stem had short internodes. It has sequential fruiting behavior and fruit sets at closed successions. Fruits have angled peduncle and vary in skin colour. It bears monoecious flowers, but androecious mutant has also been reported. Anthesis takes place early in the morning and flowers are pollinated by honeybees.

A well-drained loamy soil, rich in organic matter with proper drainage is suitable for its cultivation. It can be grown between September to April in Uttar Pradesh, Punjab, Haryana and Rajasthan, and March to June in hilly states. It can be raised throughout India under temperatures variation from 18-30°C. It fits well in multiple cropping systems. The seed requirement for 1 hectare area is 3-5 kg. Frequent irrigation at shorter interval is required. Marketable fruit picking starts at 45-50 days after sowing and can be picked 10-15 days after setting. Fruits are perishable in nature. Some of the major limiting factors in production of this crop are diseases, caused by mosaic and tospo viruses. Disease resistance has been a challenge in cultivar development mainly because resistance against major diseases are present geurally in wild species and transfer of resistance from wild to cultivated species is not an easy task. Breeding programmes should be taken up for resistance by bridging species to overcome sterility.

Varieties

The main breeding objectives are to develop high yielding varieties/ hybrids by increasing the number of fruits per plant rather than increasing the fruit weight. The crop improvement programmes on squash is being progress at IIVR (Varanasi), PAU (Ludhiana), IARI-Regional Station (Katrain) and IIHR (Bangalore). Some of the introduced varieties have been released in India after evaluation of their adaptability to specific area/environment and several varieties have been developed as detailed below.

EARLY YELLOW PROLIFIC

This variety has been developed at IARI, Regional Station, Katrain and notified by CVRC Notification no. 440 dated 21.08.1975. It is bush type and early in flowering. Fruits are of medium size, warted and tapering towards stem end with light yellow skin turning orange-yellow on maturity. Flesh is tender when it is consumed as vegetable. It is recommended for cultivation in Punjab, H.P., and Uttaranchal and has yield potential of 250-275 q/ha.

PUNJAB CHAPPAN KADDU-1

This variety has been developed at PAU, Ludhiana and notified by CVRC Notification no. 499(E) dated 08.07.1983. It is bush type and have non-lobed leaves. Fruits are green, medium in size and disc-shaped. This is an early maturing variety, ready for first harvest in about 60 days after sowing. Single fruit weight is 40-50 g. Plants have predominantly female flower bearing tendency. It is resistance to downy mildew, tolerance to CMV, powdery mildew and pumpkin beetle under field condition. It is recommended for cultivation in Punjab, H.P., and Uttaranchal and has yield potential of 300-320 q/ha.

PATTY PAN

This variety is an introduction fom USA, recommended by IIHR, Bangalore and notified by CVRC Notification no. 867(E) dated 26.11.1986. It is erect with short internodes bearing white disc-shaped fruits with notches along the margin. Single fruit weight is 350-400 g. It has yield potential of 550-575 q/ha in 90-95 days of crop duration.

ARKA SURYAMUKHI

This variety has been developed through selection from a local collection (IIHR-79) at IIHR, Bangalore and identified for release through AICRP-VC in 2001. It has broad leaves and palmate veins without any pattern on leaf lamina. Fruits are small, round with flat end, deep orange with creamy white streaks on the rind, 1-1.5 kg of weight with orange yellow flesh. It is resistant to fruit fly. It has been recommended for cultivation in the states of Karnataka, Tamil Nadu and Kerala and has yield potential of 360-385 q/ha in about 115-120 days of crop duration.

RIDGE GOURD

Luffa acutangula
Family: Cucurbitaceae
Chromosome no.: 2n=26

Ridge gourd (ribbed gourd, four-angled gourd, angular luffa, ridge loofah, silky gourd) is an important cucurbit vegetable. It probably originated in India and is found in its wild form in the North-Western parts of India. It is grown in Malaysia, Sri Lanka, Japan and Brazil. It is very popular low-cost vegetable grown in tropical and subtropical regions. *L. acutangula* includes three varieties; *viz.*, *acutangula,* the domesticated; *amara,* a wild or feral variety with small, bitter fruits, which is mostly found in India; and *forskalii*, a wild or feral taxon found in Yemen, the relationship of these is uncertain. Other related species are *L. operculate*, *L. graveolens*, *L. umbellate*, *L. achinata*, *L. pentandra*, *L. gigante*, *L. scabra* and *L. narylandica*. In India, it is cultivated in summer/rainy season. The states cultivating ridge gourd in India are Uttar Pradesh, Madhya Pradesh, Bihar, West Bengal, Delhi, Haryana, Rajsthan, Maharastra, Karnataka and Assam. Fruits are 15-40 cm long, club-shaped with ribbed surface. The edible part is fleshy and many seeded pepo. Flowers are born in raceme. Ridge gourd fruits contain 0.5% protein, 3.4% carbohydrates and 18 mg vitamin C in 100 g edible parts.

Anthesis occurs between 17.00-20.00 h, anther dehiscence takes place between 17.00-20.00 h. Pollen fertility is the maximum on the day of anthesis and lasts till 2-3 days after anthesis in winter and one day in the rainy season. Stigma remains receptive 6 h before to 84 h after anthesis. A variety of sex forms, *viz.*, monoecious, androecious, gynoecious, gynomonoecious, and andromonoecious and hermaphrodite are found in ridge gourd. Leaves are shallowly lobbed and larger without white patches. Male flowers are smaller and the ovary is disc-shaped, angled and 10-ribbed fruits, seeds are black and pitted.

A well-drained loamy soil with proper drainage is considered best. It requires warm season for maximum production. The optimum temperature for proper growth and development is 25-27°C. The sowing time in northern plains is January to February for summer crop and June to July for rainy season crops. In the southern part of the country, it can be cultivated round the year due to mild temperature during winter season. April to June is the best sowing time at hilly regions of northern plains. The seed requirement for one ha area is 3.5-5 kg. Irrigation at shorter interval is required during summer season, whereas, in rainy season it depend upon the rains. The marketable maturity of fruits depend upon the variety and growing season. The fruits are ready for harvesting after 5-7 days of fruit setting. Downy mildew, powdery mildew and mosaic virus are the main diseases causing low productivity. The objectives of ridge gourd breeding are earliness, fruits cylindrical, less-fibrous, non-bitter and possessing resistance against biotic and abiotic stresses.

Varieties

The important breeding methods utilized to evolve the varieties are inbreeding and selection, hybridization followed by selection during segregating generations, heterosis and resistant breeding. The crop improvement programmes on ridge gourd is being progress at IIVR (Varanasi), KAU (Vallanikkara), HARP (Ranchi), TNAU (Coimbatore) and IIHR (Bangalore) (for photographs see plate 21).

CO-1

This variety has been developed at TNAU, Coimbatore, and notified by CVRC Notification no. 13 dated 19.12.1978. It is prolific bearing have 10-12 fruits/vine (3-4 kg). Fruits are 60 to 70 cm long with 3.0 cm diameter. It has yield potential of 210-150 q/ha in a crop duration of 125 days.

PUNJAB SADABAHAR

This variety has been developed at PAU, Ludhiana, and notified by CVRC Notification nos. 13 and 563(E) dated 19.12.1978 and 30.08.1991. Its fruits are long, 3-5 cm thick, straight, green and tender. Best sowing time of this variety is May to July for rainy season crop and it has yield potential of 100-125 q/ha.

PKM-1

This variety has been developed through mutation breeding (genotype H-160) at TNAU, Coimbatore and notified by CVRC Notification no. 596(E) dated 13.08.1984. Fruits are dark green 60-70 cm long with shallow grooves characterised by very broad terminal than the basal end. It is tolerant to pumpkin beetle, fruit fly and leaf spot and has yield potential of 250-280 q/ha in 160 days crop duration.

CO-2

This variety is a selection from a germplasm at TNAU, Coimbatore and notified by CVRC Notification no. 540(E) dated 24.07.1985. Each vine on an average produces 8 to 10 fruits weighing 9 to 10 kg. Fruits are green, very long (10.8-1.0 m) and fleshy. It has yield potential of 250-270 q/ha in 140-150 days of crop duration.

ARKA SUMEET

This variety has been developed at IIHR, Bangalore and notified by CVRC Notification no. 664(E) dated 11.05.2005. Its vines are of medium length (4-5 m), and vigorous. Fruits are 55 cm long with 2.5 cm girth with prominent ridges and delicate aroma. It takes 52 days for first picking. Average fruit weight is 380 g, crop duration 120 days. Fruits are lush green, cylindrical with slow seed maturity. It is suitable for cultivation in Karnataka and has yield potential of 200-250 q/ha in 140-150 days of crop duration.

ARKA SUJAT

This variety has been developed at IIHR, Bangalore and notified by CVRC Notification no. 2035(E) dated 28.11.2006. Fruits are lush green and tender, medium long (35-45 cm) with an average fruit weight of 350 g. The variety has good transport and keeping quality. It is moderately resistant to downy mildew and has yield potential of 625-640 q/ha in 100-120 days of crop duration.

SWARNA MANJARI

This variety has been developed at HARP, Ranchi and notified by CVRC Notification no. 2035(E) dated 28.11.2006. Fruits are elongated, medium-sized, highly ridged and green with soft less-fiberous pulp. It is tolerant to powdery mildew. First picking starts 65-70 days after sowing. It has yield potential of 180-200 q/ha in 140-150 days of crop duration.

SWARNA UPHAR

This variety has been developed through pedigree selection at HARP, Ranchi and notified by CVRC Notification no. 2035(E) dated 28.11.2006. Fruits are elongate, medium-sized (200 g), weak ridge with soft pulp and less fiber. First picking starts at 65-70 days after sowing. It has yield potential of 280-310 q/ha in 140-150 days of crop duration.

DEEPTHI

This variety has been developed at KAU, Vellanikkara and notified by CVRC Notification no. 597(E) dated 24.04.2006. It has green ovate intermediate-sized leaves. First harvest can be done in 63 days. Fruits are green, medium-sized with finely wrinkled surface. Average fruit weight is 165 g, 23 cm long and 14.9 cm girth. It is resistance to mosaic and downey mildew under field conditions and has yield potential of 300-325 q/h.

PANT TORAI-1 (PRG-7)

This variety is a pure line selection from an indigenous germplasm at GBPUA&T, Pantnagar, and it was identified for release through AICRP-VC in 2001. The main shoot is ~5 m long. Fruits are 15-20 cm long and club-shaped. It takes about 65 days to first harvest. The variety is suitable for rainy season. It is recommeneded for cultivation in Madhya Pradesh and Maharashtra and has yield potential of 100-120 q/ha.

WATERMELON

Citrullus lanatus
Family: Cucurbitaceae
Chromosome no.: 2n=22

Watermelon is a well-known warm season vegetable cultivated through out the world mainly valued as a dessert. It is commonly called as *tarbooz*, and grown in all states of India. However, it is a commercially riverbed crop in Uttar Pradesh, Gujarat, Maharashtra, and Andhra Pradesh. Most of the varieties are monoecious but few have andromonoecious form. Watermelon is considered to be indigenous to tropical Africa. Cultivation of watermelon began in the ancient Egypt and India. The cultivated forms show much variability in central Asia, Iran, Pakistan, and India.

Watermelon is a common man's fruit, its juice is delicious, nourishing and provides a cooling effect in summer season. Unripe fruits are cooked as a vegetable in some parts of India, and seeds after roasting and shelling are used as a snack. The fruit per 100 g fresh weight contain 95% water, 0.3% protein, 0.3% minerals, and 3.3% carbohydrates, and is also a rich source of iron. Watermelon is purgative, anthelmintic, antipyretic, and carminative in action. It cures a wide range of ailments like enlargement of spleen, ulcers, asthma, bronchitis, jaundice, anaemia, dyspepsia and constipation.

Staminate flowers usually appear before the pistillate ones. Anthesis is observed between 6.00 A.M. to 7.30 A.M. with a peak between 6.30 and 7.0 A.M. Anther dehiscence generally starts 1 h before anthesis and continues until 6.30 A.M. At high temperatures, stigmatic fluid starts drying and the stigma becomes non-receptive. Natural cross pollination in watermelon varies between 85 and 90%.

Watermelon can be grown in well-drained sandy loam soil rich in organic matter, and slightly acidic in reaction (pH 5.5-7.0). It is essentially a warm season crop grown in subtropical and hot arid regions. Soil temperature between 16° and 33°C with optimum 21.5°C is the best for its seed germination, while for better pollination, fertilization and fruit setting, the temperature should be between 25° and 30°C. Plants are very sensitive to low temperature and frost. For better fruit quality, there should be plenty of sunlight and high temperature at the time of ripening.

In north India, early sowing is generally done in November in riverbeds, however, under general field conditions, it is sown in 2nd fortnight of February, whereas in northeastern and western India, the best time of sowing is November to January. The optimum time for the sowing of watermelon in West Bengal is middle of November. In southern and central parts of India, sowing is done in December-January. The rainy season crop in Rajasthan is sown in the month of August-September. In northern Indian hills, sowing season starts from April-May.

The seed rate is about 2-4 kg/ha at the recommended plant spacing, keeping 3000-8500 plants per ha. Soil moisture stress during pre-flowering, flowering or fruit development stages drastically reduce the yield. Irrigation once in 7 to 10 days may be required depending upon the soil type, organic matter content and weather conditions. Watermelon is a non-climacteric fruit, which ripens only on vine. *Fusarium* wilt and watermelon bud necrosis are the major diseases affecting the crop production, while mite is an important pest.

The adoptability and popularity of hybrids can be made more viable by selecting parental lines, which are suitable for different environments and have biotic and/or abiotic stress resistance. The major objectives in watermelon breeding is to develop the varieties/ hybrids with uniform quality and appearance, dark fruit colour, dwarf plant types, prolong shelf-life, thick rind (desirable for long distance transport), deep pink or red crisp flesh, less seeds of small size, linear arrangement of seed and high early yield having a balance between number and size (medium 3-4 kg preferred) and resistant to biotic and abiotic stresses.

Varieties

Most of the varieties in India are developed following the methods like introduction, selection, hybridization, heterosis, backcross breeding and resistant breeding. The crop improvent of watermelon in India is being carried out at IIHR (Bangalore), RAU-ARS (Durgapura), and IARI (New Delhi). Some of the introduced varieties have also been released in India after evaluation of their adaptability to a specific area/environment (for photographs see plate 22).

SUGAR BABY

This is a popular American variety, introduced and recommended by IARI, New Delhi, and notified by CVRC Notification nos. 440, and 1004 dated 21.08.1975, and 23.03.1978. It has medium long vines. Fruits are round with black rind, deep pink flesh with 11-13% TSS, brown small seeds and each weighing 2-5 kg. It has been recommended for cultivation in Chhattisgarh, Orissa, A.P., M.P., Maharashtra, Karnataka, Tamil Nadu and Kerala and has yield potential of 150-170 q/ha in 85 days of crop duration.

DURGAPURA KESAR

This variety has been developed at RAU-ARS, Durgapura, and notified by CVRC Notification no. 13 dated 19.12.1978. Fruit weight is 4-5 kg having green skin with stripes. The flesh is yellow and moderately sweet. Seeds are large and yellow. It has yield potential of 400-450 q/ha.

DURGAPURA MEETHA

This variety has been developed at RAU-ARS, Durgapura, and notified by CVRC Notification no. 13 dated 19.12.1978. Fruits are round with light green and thick rind, good keeping quality, each weighing 7 to 8 kg, flesh dark with 11% TSS, seeds with black tip and margin. It has been recommended for the cultivation in Punjab, U.P., Bihar, Rahasthan, Gujarat, Haryana, Delhi, M.P., Karnataka, Tamil Nadu and Kerala and has yield potential of 480-500 q/ha in 125 days of crop duration.

SIPPER

This American variety is an introduction, recommended by PAU, Ludhiana and notified by CVRC Notification no. 13 dated 19.12.1978. Its vines are long and vigorous, fruits are medium to large weighing 6-10 kg, round, with characteristic tripartite blossom end. Flesh is red with moderate sweetness, TSS 8-9%. Rind is thick and dark green. Seeds are uniformly light brown. It has yield potential of 300-320 q/ha.

ARKA JYOTI

This hybrid has been developed through heterosis breeding (IIHR-20 × Crimson Sweet) at IIHR, Bangalore, and notified by CVRC Notification no. 867(E) dated 26.11.1986. Fruits are round to oval, dark green with blue angular stripes, each weighing 5-6 kg, flesh is bright crimson, granular texture with 12-13% TSS having good keeping and transport quality. First picking starts at 90 days of sowing. It has yield potential of 700-720 q/ha.

ARKA MANIK

Arka Manik is a multiple disease resistant variety developed through hybridization (IIHR-21 × Crimson Sweet) followed by pedigree selection at IIHR, Bangalore, and notified by CVRC Notification no. 386(E) dated 15.05.1990. It has angular stem and light green deeply-lobed foliage. Its fruits are round-oval with light green rind and dark green stripes, each weighing ~6 kg, flesh is deep crimson, granular texture, pleasant aroma, TSS 12-15% with few small seeds arranged in rows, seeds are small with dark brown specks on the testa. It has good storage and transport quality. It is resistance to powdery mildew, downy mildew and anthrancnose. It has been recommended for cultivation in Punjab, U.P., Bihar, M.P., Maharashtra, Karnataka, Tamil Nadu and Kerala and has yield potential of 600-610 q/ha in 110-115 days of crop duration.

DURGAPURA LAL (RW-177-3)

This hybrid has been developed through heterosis breeding (Sugar Baby × KS-3566) at RAU-ARS, Durgapura, and notified by CVRC Notification no. 642(E) dated 31.05.2004. Fruits each of 4-5 kg are dark green with dark thin lining, round and juicy having hard rind, dark red crisp flesh and 10-11% TSS. First picking starts at 105-110 days after sowing. It is moderately resistant to blight and bud necrosis under field condition and has yield potential of 430-450 q/ha.

SPONG GOURD

Luffa cylindrica
Family: Cucurbitaceae
Chromosome no.: 2n=26

Sponge gourd is commonly known as smooth gourd, rag gourd, vegetable sponge, loofah or ghia tori. *Luffa* is essentially old world genus and has a long history of cultivation in the Asian and African countries. Luffa probably originated in India and its wild forms are available in North-West regions of India. It is grown in Malaysia, Sri Lanka, Japan and Brazil. It is an important summer season vegetable, cultivated on commercial scale and in kitchen gardens. It can also be successfully grown during rainy season. Immature fruits are consumed as cooked vegetable. In eastern states of India, tender leaves and growing tips are used as a pot herb. Phylogenetic analyses of *Luffa* indicate that *L. cylindrica* and *L. acutangula* are related and constitute a line of evolution distinct from that of the other old world and new world species. *L. cylindrica* is known in both the wild state (var. *leiocarpa*) and as a domesticated (var. *cylindrica*) form. Wild populations have smaller and bitter fruits. Luffah occur from Burma to Philippines and Australia suggesting that domestication may have occurred in Southeast Asia rather than India as commonly proposed.

Fruits are fleshy and many seeded pepo, and flowers appear as raceme. Other related species are *L. operculate*, *L. graveolens*, *L. umbellate*, *L. achinata*, *L. pentandra*, *L. gigante*, *L. scabra* and *L. narylandica*. The sex forms in sponge gourd are monoecious, androecious, gynoecious gynomonoecious, and andromonoecious and hermaphrodite. Flower anthesis occurs early in the morning between 4-8 h. Anther dehiscence also takes place early in the morning. Pollen fertility is maximum on the day of anthesis and remain till 2-3 days after anthesis in winter and one-and-a-half days in the rainy season. Stigma is receptive 6 h before to 84 h after anthesis. Generally, monoecious form is seen in sponge gourd. Leaves are deeply lobbed and small with white patches. Male flowers are larger, ovary is cylindrical and tomentose. Its fruits are white flat and smooth with black seeds. Sponge gourd closely resembles to ridged gourd.

A well-drained loamy soil is considered the best for *Luffah* cultivation. *Luffah* requires long warm season for maximum production. Optimum temperature for proper growth and development is 25-27°C. The sowing time in northern plains is January to February for summer crop and June to July for rainy season crops. In the southern part of the country, it can be cultivated round the year due to mild temperature during winter season. April to June is the best sowing time in hilly regions of northern plains. Seed requirement for one hectare area is 2.5-3.5 kg. Irrigation at shorter interval is required during summer season, whereas in rainy season, it depend upon the rains. Fruits are harvested at right edible maturity stage. The marketable maturity of fruits is depending upon the varieties and growing season. The fruits ready for harvesting after 5-7 days of fruit setting. Picking should be done at every 2-3 days interval. Downy mildew, powdery mildew and mosaic virus are the main diseases causing low productivity.

Varieties

The main breeding objectives in sponge gourd breeding are cylindrical, tender and non-bitter fruits. The female to male sex ratio should be minimum with maximum number of fruits leads high yield. The genetic improvement of this crop is in progress at IARI (New Delhi), HARP (Ranchi), AAU (Anand), CSAUA&T (Kanpur) and MPKV (Rahuri). Several varieties have been developed and described in detail as under (for photographs see plate 23).

PUSA CHIKNI

Pusa Chikni is an early variety developed at IARI, New Delhi and notified by CVRC Notification no. 440 dated 21.08.1975. It is suitable for cultivation in rainy and spring summer seasons. Fruits are smooth, dark green, and cylindrical. Single plant bears 15-17 fruits. Flowering starts in 60 days after sowing. It has yield potential of 200-225 q/ha.

KALYANPUR HARI CHIKNI

This early variety has been developed at CSAUA&T, Kanpur and notified by CVRC Notification no. 2(E) dated 03.01.1983. It bears long, smooth, dark-green fruits with pointed distal end. It is suitable for rainy season and has yield potential of 180 -200 q/ha.

PUSA SUPRIYA

This variety has been developed at IARI, New Delhi and notified by

CVRC Notification no. 98(E) dated 08.02.1997. Its vines are 2.5-2.8 cm long with 4-5 branches. Leaves are dark green, smaller in size with six lobed petiole, tip pointed and simple. Fruits are pale green, smooth, 15-20 cm long, straight and slightly curved at the stem end, pointed distal end, non-hairy, flesh tender, suitable for spring summer and Kharif season. Fruits become ready for picking at 50-55 days after sowing in spring-summer and 44-48 days after sowing in kharif season. It is recommeneded for cultivation in Punjab, U.P., Bihar, Jharkhand, Rajasthan, Gujarat, Haryana and Delhi and has yield potential of 130-140 q/ha.

PHULE PRAJAKTA (GFE-SMG-108)

This variety has been developed through selection from a germplasm (GFE-SMG-108) at MPKV, Rahurui and notified by CVRC Notification no. 348(E) dated 19.04.2001. Fruits are medium-sized, green, straight, slender, tapering at ends and pure white flesh. It is highly tolerant to downy mildew, fruit fly and red pumpkin beetle, and moderatly resistant to wilt and bean mosaic virus. It has yield potential of 130-150 q/ha in 145-150 days of crop duration.

PUSA SNEHA

This variety has been developed at IARI, New Delhi and notified by CVRC Notification no. 642(E) dated 31.05.2004. Its vines are 2.5-3 m long with 4-5 branches. Its leaves are dark green, small in size with 6 lobed margin, non-hairy, smooth having long petioles. Fruits are attractive, dark green, 20-25 cm long, straight with long, smooth and non-hairy peduncle, flesh tender, skin is hard suitable for long distance transportation. It is suitable for growing in both spring-summer and kharif seasons. it becomes ready for first picking at 50-55 days after sowing in spring summer and 40-45 days in kharif season and has yield potential of 200-230 q/ha.

GUJARAT SPONGE GOURD-1

This variety has been developed from a germplasm at GAU, Anand and notified by CVRC Notification no. 597(E) dated 25.4.2006. Its fruits are medium-sized and light green. It is moderately tolerant to downey mildew and leaf miner. It has been recommended for cultivation in Madhya Pradesh and Maharashtra and has yield potential of 120-140 q/ha.

SWARNA PRABHA (CHSG-1)

This variety has been developed through pure line selection at HARP, Ranchi and notified by CVRC Notification no. 597(E) dated 25.4.2006. Its fruits are medium in length (20-25 cm), each weighing 150-200 g, light greenish, pulp soft and contain less fibre. It is tolerant to powdery and downy mildew and resistant to leaf miner under field conditions. First harvest can be taken at 70-75 days after planting. It is recommended for July-September and February-April sowing in Uttar Pradesh, Jharkhand and Bihar and has yield potential of 200-250 q/ha.

AZAD TORAI CHIKNI-1

This variety has been developed at CSAUA&T, Kanpur and notified by CVRC Notification no. 597(E) dated 25.4.2006. It is an early maturing variety having smooth green fruits. This variety is suitable for rainy season and has yield potential of 180-200 q/ha.

AZAD TORAI-2 (KSG-14)

This short duration high yielding variety has been developed at CSAUA&T, Kanpur and identified for release through AICRP-VC in 2006. Its fruits compact, attractive green, and rich in nutrients. It is recommeneded for cultivation in Punjab, U.P., Bihar and Jharkhand and has yield potential of 150-200 q/ha.

ASH GOURD

Benincasa hispida
Family: Cucurbitaceae
Chromosome no.: 2n=24

Ash gourd belongs to genus *Benincasa* of family Cucurbitaceae with single cultivated species *hispida* syn. *cerifera*. It is also known as wax gourd, white gourd, white pumpkin, hairy melon, Chinese preserving melon, and winter melon. Ash gourd is normally grown is South East Asia, China and India. In India, it is grown in Punjab, Uttar Pradesh, Bihar, West Bengal, north-eastern hill states, Orissa, Kerala and Tamil Nadu on commercial scale. Ash gourd is believed to have originated in Java and Japan. The diversity of cultivar in China suggests that this crop might be indigenous to southern China.

Its vines are thick, furrowed (angular), hispid with coarse hair with cylindrical and hairy tendrils. Leaves are irregular 5-7 lobed, reniform-orbicular, deeply cordate, scabrous, lower rigidly hispid and sinuate margin. The golden yellow solitary flowers in leaf axil are monoecious, and the calyx teethed at early stage are often narrow and serrated. Fruits are fleshy, succulent and densely hairy at immature stage but thickly deposited with white easily removable waxy layer. Flesh is white and generally spongy. Seeds are numerous, flat, smooth with narrow base, white, yellow or pale brown. Seed size is 1 × 0.5 cm containing yellow oil.

Mature fruits of ash gourd are used in making confectionery and immature fruit are used as culinary vegetable. Specific cultivars or land races are grown for vegetable purpose or for preparation of *sweets* and *bari*. The ash gourd enzymes may have potential as an alternative for calf rennet in cheese manufacture. Extract of ash gourd has acid neutralizing property and recommended in *Ayurveda* for management of peptic ulcer. Fruit juice is used for treating a range of ailments, including insanity epilepsy. It is used for improvement of musculature and also for blood sugar patient. It is also considered good for people suffering from nervousness and disability. The fruits contain tonic, nutritive and medicinal properties. Cucurbitacin from ash gourd had protective effect on kidney damages caused by mercuric chloride.

The crop can be grown on a variety of soils, ranging from sandy loam to clay loam rich in organic matter with good drainage. It can be grown in soils with a pH ranging from 5.0 to 7.5. In northern plains of India, the best time of sowing for summer crop is second fortnight of February, while suitable time for the sowing of rainy season crop is June-July in northwestern India and June to August in south India. Suitable sowing time in hills of northern India is April-May. In northern plains, if irrigation facility is available sowing may be done in the month of April and crop should be raised at bower but the height of bower should not be more than 4.0-4.5 feet otherwise the fruit may detach from the vine due to load. The seed rate would be about 2.5-3.0 kg/ha at recommended plant spacing, keeping population density of 5000-6000 plants per hectare.

Ash gourd has extensive root system and responds well to irrigation. In rainy season, no irrigation is required however, in summer season, light irrigation is given just after seeds sowing to facilitate germination, and subsequent irrigations are given at 8-10 days interval. The crop usually matures in about 90-140 days after seed sowing. Development of a thick layer of white wax is the index for judging the right stage for harvesting of fruits at full maturity. Small, solid, green, hairy, immature fruits, which are usually harvested after 10-12 days of anthesis, are the best for culinary purpose. Anthracnose and fruit rot are the main diseases causing low productivity. Fruit fly and leaf miner are becoming limiting factor to get the good yield.

Varieties

The main breeding objective of ash gourd improvement is high flesh recovery varieties with linear seed arrangements. The improvement of ash gourd in India is in progress at IIVR (Varanasi), IARI (New Delhi), KAU (Vellanikkara), PAU (Ludhiana), GBPUA&T (Pantnagar) and TNAU, (Coimbatore). Several varieties have been developed and described in detail as under (for photographs see plate 24).

CO-2

This variety has been developed at TNAU, Coimbatore, and notified by CVRC Notification no. 596(E) dated 13.08.1984. Fruits are small, with average weight of 3 kg, long spherical with green flesh. It has yield potential of 230-250 q/ha in 120 days of crop duration.

PUSA UJJWAL (DAG-1)

This variety has been developed at IARI, New Delhi, and notified by CVRC Notification no. 642(E) dated 31.05.2004. Fruits oblong, ellipsoid, rind greenish white while flesh is white with avarage fruit weight of ~7.0 kg. Its fruits are ideal for long distance transportation. It is recommeneded for cultivation in the states of Karnataka, Tamil Nadu and Kerala and has yield potential of 480-500 q/ha (kharif season) and 410-425 q/ha (summer season).

KASHI UJAWAL (IVAG-90)

This variety has been developed at IIVR, Varanasi, and notified by CVRC Notification no. 2035(E), dated 28.11.2006. Fruits are globular, each weighing 10-12 kg and less-seeded. It is recommeneded for cultivation in Punjab, U.P., Bihar, Jharkhand, Karnataka, Tamil Nadu and Kerala and has yield potential of 600-610 q/ha in 130-140 days of crop duration.

KASHI DHAWAL (IVAG-502)

This variety has been developed from a local collection at IIVR, Varanasi, and notified by CVRC Notification no. 597(E) dated 25.4.2006. Its vine length is 7.5-8 cm bearing oblong fruits each of 11-12 kg. Flesh is white with 8.5-8.7 cm thickness and seed arrangement is linear, crop duration is 120 days. Due to high flesh recovery this variety is suitable for preparation of *Petha* sweets. It is recommeneded for cultivation in the states of eastern India. The variety has yield potential of 580-600 q/ha.

POINTED GOURD

Trichosanthes dioica
Family: Cucurbitaceae
Chromosome no.: 2n=22

Pointed gourd commonly known as Parwal, Parmal, Parwar, Parora, Patal, Patola, etc., is one of the most nutritive cucurbit vegetables, and holds a coveted position in the Indian market during summer and rainy seasons. It is a perennial crop, and is highly accepted due to its availability for eight months (February-September) in a year. *T. dioica* is morphologically distinct from the other cucurbitaceous species due to its well-established dioecism, perennial nature and vegetative means of propagation. The genus *Trichosanthes* is considered to be of Indo-Malayan origin. It is a large genus with 44 species, of which 22 are found in India. Although, the Bengal-Assam area in India is considered as its primary center of origin, its wild forms (*T. japonica* and *T. multiloba*) are found throughout the northern India. A large number of local varieties are popular with the growers in their respective areas.

Pointed gourd fruits are generally dark green, small, oval or light green, long, pointed at both end without striped or black irregular striping, round or regular striped having quite long or pale yellow are popular. Pointed gourd is extensively grown in Eastern parts of India (Uttar Pradesh, Bihar, Assam and West Bengal). It is also cultivated in Orissa, Madhya Pradesh, Maharashtra and Gujrat. In India, this crop is cultivated in about 14,000 ha in Bihar and in ~10,000 ha in UP. Fruits of pointed gourd are rich in protein and vitamin A. Many reports indicate its role in circulatory system, especially in lowering blood sugar and serum triglycerides. Fruits of *T. dioica* contain ample quantities of elements like Mg (9 mg), Na (2.6 mg), K (83 mg), Cu (1.1 mg), S (17 mg) and Cl (4 mg); all the values are per 100 g of edible part. Fruits are easily digestible and diuretic in nature; they are also known to have anti-ulcerous effects. In addition, fruit and other parts of the plant like leaves and tuberous root are used in the indigenous system of medicine.

Pointed gourd is propagated by vine and root cuttings, layering and tissue culture. October month is the best time for cutting when vines are mature and fruiting is completed and harvesting is over. Planting system of pointed gourd is straight, ring lunda or lachhi and moist lump method. Cutting from mature vines are planted in nursery/polytubes in October, after the fruiting is over. Cuttings are allowed to root in nursery or poly tubes and then planted in February in well-prepared field. Cuttings made from basal portion of the vines show ~90 percent success, while terminal cuttings show a high degree of mortality. Root cutting is least expensive method. Smaller root suckers of the nodes of creeper are uprooted in October-November and planted. Rooting initiates 30-35 days after planting. Root suckers may be transplanted in February. Vine cuttings of pointed gourd may be directory transplanted in September-October in uplands and in October in riverbeds. In West Bengal, the optimum time of transplanting (root cuttings) is second fortnight of October.

Soil should be sandy or sandy loam rich in organic matter with soil pH 6.0-7.5. After field preparation in the month of June-July (for direct seeded crop), pits having size of 45 × 45 × 45 cm are dug. Planting in pits at a distance of 3 × 3 m, is generally practiced. Pits are filled with a mixture of soil and 5 kg well-rotten farmyard manure. It is advantageous to mix half kg neem cake in each pit to avoid insect pest and disease infestation. Poor drainage and water stagnation upto 18-24 hours may lead yellowing of crop and wilting. Critical period of irrigation are vegetative growth, pre-flowering and fruit development stages. Since pointed gourd is dioecious in nature, it is essential to maintain the ratio of female and male plant to 10:1 in order to get the high yield with good quality fruits.

Varieties

The main breeding objective of pointed gourd improvement is to develope a variety with prolific bearing capacity. The improvement in India is in progress at IIVR (Varanasi), HARP (Ranchi), RAU (Pusa), BCKV (Kalyani) and NDUA&T, (Faizabad). Several varieties have been developed and described in detail as under (for photographs see plate 25).

RAJENDRA PARWAL-1

This variety has been developed at RAU, Pusa and notified by CVRC Notification no. 636(E) dated 02.09.1994. Fruit are long green with white strips and tapering at both ends. Average fruit weight is 40 g with good long distance transportation quality and tolerant to fruit fly. It is suitable for cultivation in both diara land and upland areas and has yield potential of 175-190 q/ha.

RAJENDRA PARWAL-2

This variety has been developed at RAU, Pusa and notified by CVRC Notification no. 636(E) dated 02.09.1994. Fruits are drum shaped, whitish green with very light self strip and soft. Average fruit weight is 30 g. It is tolerant to vine and fruit rot as well as fruit fly. It is suitable for cultivation in both diara and upland areas and has yield potential of 160-180 q/ha.

SWARNA REKHA

This variety has been developed through clonal selection at HARP, Ranchi, and notified by CVRC Notification no. 2035(E) dated 28.11.2006. Its fruits are elongated, striped green with soft seeds. It has been recommended for cultivation in Jharkhand, Bihar, Orissa, West Bengal and Eastern Uttar Pradesh for transplantation in February (rooted plant) and September (vine cutting) and has yield potential of 175-200 q/ha.

SWARNA ALAUKIK

This variety has been developed through clonal selection at HARP, Ranchi, and notified by CVRC Notification no. 2035(E) dated 28.11.2006. Its fruit are elongated, light greenish, suitable for sweet preparation. It has been recommended for cultivation in Jharkhand, Bihar, Orissa, West Bengal and Eastern Uttar Pradesh for transplant in February (rooted plant) and September (vine cutting) and has yield potential of 230-250 q/ha.

IVY GOURD

Coccinia indica
Family: Cucurbitaceae
Chromosome no.: 2n=24

Ivy gourd or little gourd (Kundru) is a perennial vegetable. It is available throughout the year except winter season when crop goes under dormancy due to low temperature. In winter season its vines are pruned, which results in larger number of flowers and fruits in new season. This vegetable is grown on commercial scale in Chhattisgarh region of Madhya Pradesh, West Bengal, Bihar and Karnataka.

Several local varieties are cultivated in different parts of India. Varieties are locally classified on the basis of fruits size, length, odour and patterns of strips on fruits. Farmers of Peninsular India are cultivating local varieties having smaller green fruits with white longitudinal strips. Farmers of eastern and western parts of U.P. cultivate varieties having fruits long, dark green with scattered small longitudinal white strips with rounded end. In Bihar, large, green fruits of local cultivar are very popular.

Ivy gourd is propagated by vine and root cutting. The vine cuttings could be directly propagated during September-October and transplanted in nursery. Soil should be sandy or sandy loam rich in organic matter with soil pH 6.0-7.5. After field preparation in the month of June-July (for direct planted crop), pits having size of 30 × 30 × 30 cm are dug. Planting in pits at a distance of 3 × 3 m is generally practiced. Pits are filled with a mixture of soil and 5 kg well-rotten farmyard manure. It is advantageous to mix half kg neem cake in each pit to avoid insect pest and disease infestation. Poor drainage and water stagnation upto 18-24 hours may lead yellowing of crop and wilting. Critical period of irrigation are vegetative growth, pre-flowering and fruit development stages. These crops are dioecious in nature, therefore it is essential that the ratio of female and male plant should be balanced to get the high yield with good quality fruit. The female and male ratio should be 10:1. Male plant should be transplanted in a manner that each male plant may cover 10 female plants for proper pollination.

Varieties

The main breeding objectives in ivy gourd improvement are prolific bearing varities with easy rooting properties. The improvement in India is in progress at IIVR (Varanasi), IGKV (Raipur), and KAU, (Vellanikkara). Several varieties have been developed as detailed below.

INDIRA KUNDRU-5

This variety has been developed at IGKV, Raipur, and notified by CVRC Notification no. 2035(E) dated 28.11.2006. Fruits are light green and oval (4.30 cm fruit length and 2.63 cm fruit diameter)). It is a high yielding variety, producing about 21 kg fruits per plant. This variety has yield potential of 400-420 q/ha.

INDIRA KUNDRU-35

This variety has been developed at IGKV, Raipur, and notified by CVRC Notification no. 2035(E) dated 28.11.2006. Fruits are long and light green 6.0 cm of length and 2.43 cm fruit diameter. It is a high yielding variety, producing about 22 kg fruits per plant and has yield potential of 410-450 q/ha.

SULABHA (CG-23)

This variety has been developed through clonal selection from a germplasm (CG-23) at KAU, Vellanikara, and notified by CVRC Notification no. 597(E) dated 25.04.2006. Fruits are long (9.25 cm), pale green, each weighing ~20 g. It takes 37-40 days from planting to first flowering. The first harvest can be done in 45-50 days after planting. Average fruit length is 9.25 cm. Fruits are cylindrical, slightly green with continuous striations. Leaves are typically trilobed. It produces ~1050 fruits/year and has yield potential of 400-425 q/ha.

SNAPMELON

Cucumis melo var. *momordica*
Family: Cucurbitaceae
Chromosome no.: 2n=24

Snapmelon is a native of India and is extensively grown in 19th century in northern India, where it is commonly known as *phoot* which means 'to split'. Immature fruits are cooked or pickled. The low sugared mature fruits are eaten raw. Ripe fruits invariably crack. Snapmelon germplasm has been found to be a good source of disease and insect resistance. The cultivation of snapmelon is confined in Uttar Pradesh, Rajasthan, Haryana, Punjab and Bihar. It is generally cultivated as mixture crop with maize, sorghum and pearlmeillet.

It can be grown in all type of soil, while light (sandy loam) rich in organic matter, are considered to be the best. Snapmelon is warm season crop. Temperatures between 18-38°C are suitable for proper vegetative and reproductive growth. June-July is the main sowing season in northern plains of India. Seed requirement is 4-5 kg/ha. Frequent irrigation is required during summer season and during rainy season it depends upon the rains. Fruits can be harvested about 90-100 days of sowing.

Varieties

The genetic improvement of this crop has recently been started at IARI (New Delhi) and IIVR (Varanasi). Only one variety has been developed as detailed below.

PUSA SHANDAR

This variety has been developed at IARI, New Delhi, and notified by CVRC Notification no. 642 (E) dated 31.05.2004. Fruits are oblong, cylindrical, each weighing ~700 g, skin creamy white with light pink flesh. It has yield potential of 400-425 q/ha (in summer season) and 370-380 q/ha (in rainy season).

LONGMELON

Cucmis melo var. *utilissimus* (syn. var. *flexuosus*)
Family: Cucurbitaceae
Chromosome no.: 2n=24

Longmelon (serpentmelon) is considered as a minor cucurbit and grown in pockets in India. It is commonly known as kakari, sara, kachari, etc. Fruits are generally consumed immature as salad or pickled. Longmelon probably originated in India. In India, the main states cultivating the longmelon are Rajsthan, U.P., Bihar, Delhi, Haryana and Punjab. Plants are monoecious, *i.e.*, staminate and pistillate flowers produced separately on the same plant. Anthesis takes place early in the morning and pollen dehiscence takes place same day. The duration of pollen viability is on the same day of anthesis and stigma receptivity starts before 36 hrs of flower anthesis and remains receptive after 48 hrs of anthesis. The long and tender fruits are preferred by consumer.

A well-drained loamy soil with proper drainage is considered best. It requires long warm season for maximum production. The optimum temperature for proper growth and development is 25-35°C. Sowing time in northern plains is January to February. April to June is the best sowing time in hilly regions of northern plains. The seed requirement for one hectare area is 2-3 kg. Irrigation at shorter interval is required during summer season. The marketable maturity of fruits depends upon the variety and growing season. Fruits are ready for harvesting after 7-9 days of setting. Picking should be done at alternate days. Downy mildew, powdery mildew and mosaic virus are main diseases causing low productivity.

Varieties

It is an unexploited crop from research point of view as only two varieties are grown at commercial scale. The main breeding goal in this crop is to develop a variety having tender small fruits with profuse bearing. Fruits are highly susceptible to fruit fly, therefore less-susceptible varieties should be the priority. The improvement of longmelon in India is being progress at PAU (Ludhiana), IIVR (Varanasi) and IIHR, (Bangalore). Several varieties have been developed and released as detailed below (for photographs see plate 26).

ARKA SHEETAL

This variety has been developed through selection at IIHR, Bangalore and notified by CVRC Notification no. 867(E) dated 26.11.1986. Fruits are medium long, very tender, dark green, crisp, fleshy, each weighing 90-100 g and free from bitter principle. It has yield potential of 210-250 q/ha in 100-110 days of crop duration.

PUNJAB LONG MELON-1 (H-10)

This variety has been developed through selection at PAU, Ludhiana, and notified by CVRC Notification no. 115(E) dated 10.02.1996. Fruits are long, thin and light green. It has yield potential of 200-220 q/ha.

ROUNDMELON

Praecitrullus fistulosus
Family: Cucurbitaceae
Chromosome no.: 2n=24

Roundmelon or tinda is an important summer vegetable cultivated extensively in north India. It is believed to have originated in India. The genetic variability in morphological traits is concentrated in North-western India (western U.P., Punjab and Haryana). Immature fruits are cooked as vegetable, fruits contain 1.4% protein, 0.4% fat, 3.4% carbohydrates, 13 mg carotene and 18 mg vitamin C per 100 g fresh weight. Fruits are also considered good for treating cough and improving blood circulation.

It is monoecious, *i.e.*, staminate and pistillate flowers produced separately on the same plant. The flower anthesis takes place early in the morning (5-7 hrs), whereas pollen dehiscence takes place at the same day. The duration of pollen viability is on the same day of anthesis and stigma receptivity starts before 20 hrs of flower anthesis and remains receptive after 36 hrs of anthesis. The round and tender fruits are preferred by the consumers. Fruits remain tender and non-fibrous at edible maturity and green in colour are highly acceptable.

The crop being hardy can be grown on a wide range of soils, however, sandy loam or loam soil rich in organic matter with pH of 6.5-7.5 is ideal for its cultivation. Round melon being warm season vegetable crop, is grown in spring-summer and in rainy season. It can be successfully grown in regions, where the temperature ranges from 20 to 35°C. A temperature of 27-30°C is the most optimal for its seed germination, while 30-35°C is the best for fruits setting. The seeds are sown in the month of February and March for spring-summer crop and in June and July for rainy season crop, however in southern India, two crops can be taken, spring-summer crop is sown in the month of December-January and rainy season crop in June-July. In hilly regions of north India, sowing is done in May-June and could be continued up to July. For one hectare area, ~3.5 kg seeds are sufficient.

Varieties

The improvement programme in this crop is still its infancy. The main breeding objectives are to develop high yielding varieties with good quality. The improvement of roundmelon in India is being progress at PAU (Ludhiana), and IIHR, (Bangalore). None of the roundmelon variety has been tested under the AICRP-VC indicating limited research. However, few varieties have been developed through selection as detailed below (for photographs see plate 27).

PUNJAB TINDA (S-48)

This variety is a selection from a local cultivar at PAU, Ludhiana and notified by CVRC Notification no. 13 dated 19.12.1978. Its plants are vigorous, bearing 8-10 fruits per vine. Fruits are medium in size with shining and pubescent surface. Flesh is white, soft, less-seeded and has good cooking quality. It takes about 60 days from sowing to marketable maturity. It has yield potential of 45-50 q/ha.

ARKA TINDA

This variety has been developed by hybridization (T-3-4 × T-8-2) followed by pedigree selection at IIHR, Bangalore and notified by CVRC Notification no. 867(E) dated 26.11.1986. It is an early summer season variety. Fruits are round, light green with soft hair and tender flesh at marketable stage. It is tolerant to fruit fly and has yield potential of 95-100 q/ha in 90-100 days.

HISAR TINDA

This variety has been developed at HAU, Hisar, and notified by CVRC Notification no. 597(E) dated 25.04.2006. It is an early maturing and high yielding variety. Fruits round, medium in size and tender. It has yield potential of 75-100 q/ha.

SNAKE GOURD

Trichosanthes cucumerina (syn. *T. anguina*)
Family: Cucurbitaceae
Chromosome no.: 2n=22

Snake gourd is grown in peninsular and eastern parts of India and more predominantly in south India. It is probably a native of India, commonly known as chichinda, pallakaya, pattakaya and padavalakayi. It is a popular vegetable in South India where immature fruit is an ingredient in curry preparations. It is cultivated throughout the year. The immature fruits are boiled and eaten. Its fruits have a fairly good nutritive value and also used to improve appetite as tonic and stomachic and cure biliousness. Occasionally, shoots and tender leaves are also used as vegetable, root and seeds possess anti-helmethic properties and are also used in treatment of diarrhea, bronchitis and fever. High humidity favours growth and development of plants and fruit formation. Fruits are narrow, cylindrical and pointed at both the ends. Plants are monoecious, annual, herbaceous and climber. Female flowers are solitary, sessile with long narrow and densely hairy ovary. The male inflorescence has long, cylinderical with highly pubescent stalk, 10-20 cm long with flowers in group of 5 or more. Fruit length varies from 30 to 150 cm.

A well-drained loam soil with proper drainage is considered to be the best. It requires warm season for maximum production. The optimum temperature for proper growth and development is 25-30°C. The sowing time in northern plains is January to February for summer season and June to July for rainy season crops. In southern part of India, due to mild temperature during winter season snak gourd can be cultivated round the year. April to June is the best sowing time at hilly regions of northern plains. The seed requirement for one hectare area is 5-6 kg. Irrigation at shorter interval is required during summer season. Marketable maturity of fruits depends upon the variety and growing season. Fruits usually become ready for harvesting after 7-9 days of fruit set. Picking should be done at 3-4 days interval. Downy and powdery mildew are the main diseases causing low productivity.

Varieties

The breeding methods used for improvement of snake gourd are selection, hybridization followed by selection, and resistance breeding. The major objectives in genetic improvement of this crop are earliness, low female male ratio, attractive fruit colour and shape, more number of fruits and resistance against diseases and pest. The improvement of snake gourd in India is being progress at KAU (Vellanikkara), KKV (Dapoli), and TNAU, (Coimbatore). Several varieties have been developed and described in detail as under (for photographs see plate 28).

CO-1

This early maturing variety has been derived through pure line selection from a germplasm collected from Alangulam (Tirunelveli) developed at TNAU, Coimbatore and notified by CVRC Notification no. 13 dated 19.12.1978. Its fruits are long (160-180 cm), dark green with white strips and smooth whitish green flesh. Its vines are moderately vigorous and medium spreading, individual fruit weight is ~600 g. It has yield potential of 180-190 q/ha in a crop duration of 135 days.

PKM-1

This variety has been developed by induced mutation from H-375 TNAU, Coimbatore and notified by CVRC Notification no. 19(E) dated 14.01.1982. It vines are vigorous in growth. Fruits are dark green with white stripes on outer side and light green inside, average fruit weight ~700 g. Fruits are extra long (180-200 cm). This variety has yield potential of 255-270 q/ha in a crop duration of 145 days.

CO-2

This variety has been derived through pure line selection from a germplasm developed at TNAU, Coimbatore and notified by CVRC Notification no. 867(E) dated 26.11.1986. The vines are less spreading and hence seeds can be sown at a close spacing of 1.5 × 1.5 m. Fruits are short and stout, light greenish white and each weighs ~500 g. It has yield potential of 350-370 q/ha in a crop duration of 105 days.

KONKAN SWETA

This variety has been developed at KKV, Dapoli and notified by CVRC Notification no. 636(E) dated 02.09.1994. Its fruits are white and medium-long (90-100 cm), have good flesh if harvested timely, otherwise become hollow. It has yield potential of 150-200 q/ha in a crop duration of about 120-130 days.

BABY

This variety has been developed at KAU, Vellanikkara and notified by CVRC Notification no. 597(E) dated 25.04.2006. Its fruits are small, 36.5 cm in length and 22 cm width, attractive white, each weighing ~475 g and suitable for distant marketing. It has been recommended for cultivation in Kerala and has yield potential of 390-420 q/ha.

MANUSHREE

This variety has been developed at KAU, Vellanikkara and notified by CVRC Notification no. 597(E) dated 25.04.2006. It produces male flowers in the sixth and female flowers in the seventh week after seed sowing. Its fruits are attractive white, uniform, medium long (65-70 cm), each weighing ~750 g with green markings at the pedicel end. Fruits are ready for harvest in the eighth week after sowing. It has yield potential of 575-620 q/ha.

Chapter 8

Leafy Vegetables

AMARANTHUS

Amaranthus tricolor
Family: Amaranthaceae
Chromosome no.: 2n=34

Amaranthus is the most common leafy vegetable grown during summer and rainy season in India. It is one of the cheapest dark-green leafy vegetables because of low production cost and high productivity. It is a short duration crop fits well into diverse cropping systems and crop rotations. In India, amaranthus is grown all over the country, but it is more popular in south India. It is believed to be a native of India. The centres of diversity are Central and South America, India, South-East Asia and the secondary centre of diversity is Africa. The genus *Amaranthus* includes 50-60 species, cultivated for leaf (green) and grains and few are wild species. Among leafy types *A. tricolor* is the main cultivated species in India. Other cultivated species are *A. blitum* and *tristis*. It is a highly cross-pollinated crop and very little attention has been paid by the researchers towards its genetic improvement. Most of the developed varieties have been evolved through selection. The other methods of improvement are mass selection, bulk population, pedigree selection, etc. Amaranthus leaves possess excellent nutritional value and are a good source of iron, folic acid and other micronutrients.

In north India, sowing seasons are February-March and June-July, whereas in south, it can be sown round the year. Seed requirement is 1-2 kg seed/ha. To maintain sufficient soil moisture, irrigation is applied immediately after sowing. There are two types of practices of harvesting amaranthus. In one case, whole plant is pulled and sent to market after washing, while periodical cuttings made at 3-4 weeks after sowing and subsequent cuttings are made at weekly intervals.

Varieties

Amaranthus has received less attention for its improvement. Most of the varieties of amaranthus have been developed through selection. In India, the research is being carried out at TNAU (Coimbatore), and IARI (New Delhi). Few varieties have been developed and described in detail as under.

CO-2

This variety is a selection fron a local genotype (*A. tricolour*), developed at TNAU, Coimbatore and notified by CVRC Notification no. 2103 dated 12.08.1980. Its plant height is 16.3 cm in 25 days and stem is green, erect, terete, longitudinally, furrowed with moderately branching. Leaves are simple green, alternate, ovate in the initial stage and lanceolate at maturity. The leaves contain 1.3% crude fibre with 19.0 mg of iron and 20.0 mg of calcium per 100 g. Inflorescence is green having stout unbranched terminal with small flower clusters in leaf axils. Seeds are black and bolder in size (1200-1400 seeds/g) compressed with fleshy endosperm. Flowering starts at 40-45 days after sowing and seeds mature at 85-90 days. Seed yield is 8 q/ha in a crop duration of 70-80 days. It has yield potential of 200-250 q/ha.

CO-3

This variety is a selection from a local genotype (A-83), developed at TNAU, Coimbatore and notified by CVRC Notification no. 596(E) dated 13.08.1984. It belongs to *A. tricolour* var. *tristis*. Stem is succulent and leaves are green with fine petiole, erect, obviate. Seed yield is 2 q/ha. Seeds are black and medium-sized (2500-2800 seeds/g) and highly palatable. Unclipped plants flower at 35-40 days and mature at 85 to 90 days after sowing. The leaves contain 25.2 mg vitamin C per 100 g of edible part, 1.74% crude fibre, 0.8% iron and 2.48% calcium. It has yield potential of 225-250 q/ha.

ANNAPURNA

This variety is a pure line selection from a germplasm from Pauri (Uttaranchal), developed at KAU, Vellannikkara and notified by CVRC Notification no. 258(E) dated 14.05.1986. Plant height is 220-225 cm. Stem is stout generally unbranched and ridged. It has broad dark green leaves and inflorescence with long compact terminal and lateral spikelets, brackets long and pointed. Inflorescence turns greenish yellow on maturity. Seed surface is smooth with excellent popping quality (5 times to its seed size). Storage quality of seeds is excellent. Leaves are used as green vegetables due to its excellent organoleptic quality. Seeds can also be used for popping. It is suitable for early sowing from 1st or 2nd week of June. It is resistant to lodging, highly resistant to drought, leaf spot diseases and *Alternaria* spp. and hardy to frost. It has been recommended for cultivation in the states of Uttaranchal, Himachal Pradesh and North East Hill Zone and has yield potential of 250-275 q/ha in a crop duration of 145 days.

ARKA SUGUNA

This variety has been developed at IIHR, Bangalore and notified by CVRC Notification no. 2035(E) dated 28.11.2006. It provides highly succulent stem along with attractive broad and green leaves. It is ideal for multicut without becoming fibrous and becomes ready for first cutting at 24 days after sowing. It may be cut at every 10-12 days intervals (upto 90 days) and has yield potential of 270-280q/ha fresh leaves.

SPINACH BEET

Beta vulgaris
Family: Chenopodiaceae
Chromosome no.: 2n=18

Spinach beet is one of the most common leafy vegetables of tropical and sub-tropical regions. It is a rich source of vitamin A and C. It is probably a native of Indo-Chinese region. Uttar Pradesh, West Bengal, Maharastra and Gujarat are spinach growing states in India.

Spinach beet is a frost tolerant crop, generally grown during winter season. It can be grown in all types of soil with pH of ~7.0 having good fertility and drainage, however sandy loam soil is preferred. Last week of August to second week of November is the main growing season in the plains. In hill region, best sowing time is March-May. It may be grown throughout the year under mild temperatures. The requirement of seed for winter season sowing is 10-15 kg/ha, whereas, for summer season it is 25-30 kg/ha. Soaking of seeds in water before sowing facilitates the germination. Irrigation should be applied at an interval of 4-6 days during summer and 10-12 days during winter. Being a leafy vegetable, the crop should be free from weeds otherwise weeds will deteriorate the quality of produce. Generally, harvesting of leaves starts after 3-4 weeks of sowing. Subsequent cuttings should be taken at an interval of 20-25 days depending upon variety and season.

Varieties

The methods of improvement of palak include introduction and selection. Mutation breeding has also been employed for development of cultivars. In India, the genetic improvement work is undertaken at HAU (Hisar), TNAU Horticultural Research Station, (Ooty), IARI (New Delhi), and University of Udaipur (Jobner). Several varieties have been developed as detailed below.

PUNJAB GREEN

This is an early variety from PAU, Ludhiana, and notified by CVRC Notification no. 843(E) dated 21.09.1998. It is suitable for growing in September. Its leaves are uniform, green and tender with entire leaf margin. It produces seed stalk in ~75 days. It is a heavy yielder, giving 6 to 7 cuttings at 15-20 days interval. It has yield potential of 125-140 q/ha fresh leaves.

OOTY-1

This variety is a selection from local genotype at TNAU, Coimbatore, and notified by CVRC Notification no. 1092(E) dated 26.10.1999. Cutting starts 45 days after sowing and continued at 15 days intervals for a period of two years. Its leaves are green and rich in vitamin A. It can be grown throughout the year and has yield (fresh leaves) potential of 150 q/ha.

ARKA ANUPAMA

This variety has been developed by hybridization (IIHR-10 × IIHR-8) followed by pedigree selection at IIHR, Bangalore and notified by CVRC Notification no. 597(E) dated 25.4.2006. Leaves are medium large, dark green, succulent, crinkled and attractive. Its leaves are rich in chlorophyll, vitamin C and iron content. It is a late bolter with vigorous initial growth and regenerates at a faster rate. Leaves become ready for first picking after 30 days of sowing. Its cutting can be taken at every 10-12 days interval (upto 75 days). It is resistant to *Cercospora* leaf spot under field conditions and has yield potential of 410-425 q/ha green leaves in four cuttings.

Chapter 9

Others

OKRA

Abelmoschus esculentus
Family: Malvaceae
Chromosome no.: 2n=130

Okra (bhendi, lady's finger, gumbo) is one of the most widely grown vegetables from tropical to sub-tropical and warmer parts of temperate zone of the country. About 14 species are listed under genus *Abelmoschus*. It is widely cultivated as summer and rainy season crop in North India and winter crop in Gujarat, Andhra Pradesh, Karnataka and Tamil Nadu. The genus *Abelmoschus* is accepted to be of Asiatic origin but the origin of cultivated species *A. esculentus* is under dispute. India, Ethiopia, West Africa and Tropical Asia are considered to be the centres of their origin.

The distinguishing feature of the genus is the caduceus calyx where calyx, corolla, and the stamen are fused together at the base and falls off as one piece after anthesis. Flower-bud initiation takes place 22-26 days after sowing. The first flower appears 20-22 days after initiation. Flowering continues for 40-60 days. Edible fruit is formed within 3-6 days of flower opening. Stigma is the most receptive on the day of flower opening when anther dehiscence and dusting of the stigma with pollen are almost simultaneous. Anthesis is observed between 9 and 10 h and is preceded by maximum anther dehiscence between 8 and 9 h. The stigma remains receptive on the day of anthesis. Protogyny has been reported in okra, which is mainly responsible for allogamy in the crop. The flower structure in okra is homogamous in which both anthers and stigma mature simultaneously resulting in self-pollination. The extent of cross-pollination depends upon cultivar, competitive flora, insect population and season, etc.

Hot and humid climate is the best for okra cultivation. It is highly susceptible to frost and low night temperature. Optimum temperature for growth and yield ranges between 25-35°C. It can be grown on a wide range of soil, but loam soil rich in organic matter with pH 6.0-6.8 is the best. In Northern plains, Kharif crop can be sown during June-July and spring-summer crop during February-March. In southern states of India, it can be grown through out the year. April-June is suitable sowing period on hills of North India. In eastern states of India, spring summer crop can be sown during late January to early February. The seed requirement for spring-summer season crop is 18-22 kg/ha, while it is 8-10 kg/ha for *kharif* season crop. Seeds are generally sown on the ridges to insure the proper germination and good stand of crop. Sowing on ridges facilitates the irrigation in spring-summer and drainage in *kharif* season. Maximum germination has been recorded when seeds are soaked for 24 hours before germination.

Generally, pods are ready for harvesting within 4-5 days after flower anthesis. It is advisable to harvest the loads at every alternate day. For nearby markets, harvesting can be done in morning hours, while for distant market harvesting in late evening is practiced. Yellow vein mosaic virus (YVMV) followed by enation leaf curl virus (ELCV) cause maximum losses in the okra crop. Insects like Jassid, mite, shoot and fruit borer are also reported to cause severe damage.

Okra is often cross-pollinated crop, therefore, maintenance breeding is essential to maintain the purity of lines. Introduction, pure line selection, intra-specific hybridization, interspecific hybridization, back cross and mutation breeding methods are utilized for improvement of this crop.

Varieties

Improvement in yield contributing traits coupled with resistance for YVMV and ELCV are major breeding objectives in okra. The major improvement of okra is being carried out at IIVR (Varanasi), IIHR (Bangalore), IARI (New Delhi), PAU (Ludhiana), HAU (Hisar), TNAU (Coimbatore), NDUA&T (faizabad), MAU (Parbhani) and GAU (Ahamadabad). Several varieties have been developed and described in detail as under (for photographs see plate 29).

PUSA SAWANI

This variety has been developed at IARI, New Delhi and notified by CVRC Notification no. 4045, 716, 361(E), 440, 786 and 540(E) dated 24.09.1969, 20.02.1970, 30.06.1973, 30.08.1975, 02.02.1976, and 24.07.1985, respectively. It was derived from an intervarietal hybridization between IC-1542 (field resistance to YVMV) and Pusa Makhmali (a commercial cultivar). Plants are 120-180 cm tall. Fruits are smooth with 5 ridges, dark green and 18-20 cm long when fully developed. It is distinguished by the presence of a purple patch at the base of the yellow petal on both the sides. It is free from the YVMV under field conditions and is suitable for plains of northern India in virus-free period (spring-summer season). In hills, it can be sown form April to May depending on the altitude. It has been recommended for general cultivation throughout the country and has yield potential of 150-175 q/ha.

P-13

This variety has been developed at PAU, Ludhiana and notified by CVRC Notification no. 563(E), dated 30.08.1991. It is an early (fruiting starts after 45 days) and heavy yielding variety. Plants are spreading with 3-4 secondary branches bearing tender long, smooth and light green fruits at every node. It has yield potential of 250-300 q/ha.

CO-2

This hybrid has been developed through heterosis breeding (AE-180 × Pusa Sawani) at TNAU, Coimbatore and notified by CVRC Notification no. 13 dated 19.12.1978. Its fruits are fairly long (20.60 cm) with girth of 6.8 cm and each weighing 33 g. It contains 12.6% crude fibre,

15.2% starch, 19.2% protein, 2.84% fat and 50.5 mg ascorbic acid per 100 g of fruit. Fruit surface is less hairy with a better consumer appeal and market preference. This variety is not suitable for summer season cultivation because of its susceptiblity to YVMV. It has yield potential of 165-180 q/ha in 90 days of crop duration.

PUNJAB PADMINI

This variety has been derived from a cross (*A. esculentus* × *A. manihot* ssp. *manihot*) developed at PAU, Ludhiana and notified by CVRC Notification no. 499(E) dated 08.07.1983. Plants are tall with 4-5 secondary branches and grow upto 180-200 cm at 45 × 30 cm spacing. Stem, shoots, petiole and basal veins of the lower surface of the lamina, are mildy scarlet red. Stem, petioles and leaves are pubscent. First picking starts at 53-54 days after sowing. Fruits are quick growing, dark green, shining smooth, thin, 15-20 cm long, weighing 20-21 g, 5-ridged and remain tender for a longer period (3-4 days). Plants are resistance to yellow vein mosaic virus and tolerance to jassids and cotton boll worm. This variety has been recommended for cultivation during spring, summer and rainy seasons under north India conditions and during winter in the south India and has yield potential of 100-125 q/ha (green pod) and 12-15 q/ha (seed).

MDU-1

This variety has been developed through mutation breeding at TNAU, Coimbatore and notified by CVRC Notification no. 540(E) dated 24.07.1985. It is dwarf with determinate plant growth having shorter internodes and each node bears fruits. Fruits are long, light green in colour with less crude fibre content (12.3%). It takes 33 days to flowering and 43 days to first picking. Fruits are light green, 20 cm long with individual fruit weight of 30 g. It can be cultivated throughout the year under irrigated conditions. It has been recommended for cultivation in southern part of Tamil Nadu and has yield potential of 115-140 q/ha in a crop duration of ~90 days.

GUJARAT BHINDI-1

This variety is a pure line selection, has been developed at AAU, Anand and notified by CVRC Notification no. 258(E) dated 14.05.1986. Plants grow to about 60-cm height in summer season and 90 cm in rainy season. Plants are erect with purple tinge on stem. Leaves are broad, dark green with purple tinge on veins. It takes ~40 days to first flowering and 55-60 days first picking after sowing. Fruiting starts from 4-5th node. Fruits are tender with 5 ridges, 14-15 cm long and 6-7 cm girth in the centre. It has yield potential of 70-90 q/ha green pods.

SELECTION-2

This variety has been developed at IARI, New Delhi and notified by CVRC Notification no. 258(E) dated 14.05.1986. It is a derivative of the cross (Pusa Sawani × Best-1) × (Pusa Sawani × IC-7194). Plants are of average height (110 cm) with occasional branching. Flowers have purple petal. Fruits are green, long, 5-edged, tender and available for market after 50 days of sowing. It is tolerant to YVMV and has yield potential of 120-140 q/ha.

PARBHANI KRANTI

This variety has been developed at MPKV, Rahuri and notified by CVRC Notification no. 867(E) dated 26.11.1986. It has been evolved through interspecific cross between *A. esculentus* cv. Pusa Sawani and *A. manihot* (a Japanese species carrying resistance to YVMV). Its plants are tall without branching with dark green foliage. Leaves are deeply lobed with narrow leaflets in the 1/3rd portion. Fruits are dark green, smooth, tender, slender, 5 ridged with long and narrow top and stalk is funnel-shaped. Fruits remain tender even if there is a delay of 2-3 days in harvesting. First harvest starts at ~55 days after seed sowing. This variety has been recommended for general cultivation throughout the country. It gives green fruit yield of 85-90 q/ha (summer crop) and 115 q/ha (rainy season crop).

PUNJAB-7 (P-7)

This hybrid has been developed through heterosis breeding (Pusa Sawani × *A. manihot* ssp. *manihot*; PDIB-2 × PDIB-1) at PAU, Ludhiana and notified by CVRC Notification no. 471(E) dated 05.05.1988. It is medium tall, stem with short internodes and splashes of pigmentation. Leaves are deeply lobed upto the bases of the petiole and leaf margins are less serrated. Stem, leaves and petiole are sparsely hairy. Fruits are medium long, green, tender and 5-ridged.

The top of the fruit is blunt and slightly furrowed. It flowers in 45-50 days and is ready for first picking after 54 days of sowing. The first fruit is borne on the 5-6th node. Plants are resistant to YVMV. It has been recommended for cultivation throughout India and possess yield potential of 95-110 q/ha (rainy season crop) and 50-70 q/ha (spring season crop).

UTKAL GAURAV (BO-2)

This variety has been developed at OUAT, Bhubaneshwar and notified by CVRC Notification no. 636(E) dated 02.09.1994. It is tall bearing green long fruits with five ridges. Fruit length is 15 to 18 cm and girth is 5.6 to 6.5 cm. Fruit does not lose its tenderness even in delayed harvest. Picking starts at 45 to 53 days after seed sowing. It tolerant to YVMV and has yield potential of 95-110 q/ha.

VARSHA UPHAR (HRB-9-2)

This variety has been developed through hybridization (Lam Selection-1 × Parbhani Kranti) followed by pedigree selection at HAU, Hisar and notified by CVRC Notification no. 636(E) dated 02.09.1994. Plants are medium tall (90-120 cm) with 2-3 branches. Fruits are smooth, dark green with 5 ridges and are ready for first picking at 47 days of seed sowing. It is resistant to YVMV and has yield potential of 100-120 q/ha (green pod) and 15-20 q/ha (seed).

HISAR UNNAT (HRB-55)

This variety has been developed at HAU, Hisar and notified by CVRC Notification no. 98(E) dated 8.2.1997. Its height is 100-110 cm and fruits are smooth with 5 ridges, fruit length 8-9 cm and diameter 1.4 cm. It is moderately tolerant to leaf hopper and resistant to YVMV and has yield potential of 100-125 q/ha.

CO-3 (HYBRID-8)

This hybrid has been developed through heterosis breeding (Parbhani Kranti × MDU-1) at TNAU, Coimbatore and notified by CVRC Notification no. 1052(E) dated 26.12.1999. Its fruits are dark green, medium in size and suitable for fresh market. Plants exhibits moderate resistance to YVMV. It is suitable for growing in *kharif* as well as summer and has yield potential of 160-180 q/ha.

PHULE KIRTI (RHROH-4)

This variety has been developed through selection from a germplasm at MPKV, Rahuri and notified by CVRC Notification no. 2035(E) dated 28.11.2006. It has green stem with red patches, fruits attractive green in colour with 7-9 cm fruit length. It has yield potential of 370-390 q/ha in 80 days of crop duration.

SHEETALA UPHAR (DVR-1)

This hybrid has been developed through heterosis breeding (PDIB-2 × PDIB-1) at IIVR, Varanasi and notified by CVRC Notification no. 1135(E) dated 15.11.2001. It is green fruited, resistant to YVMV as well as ELCV and has yield potential of 265-280 q/ha.

SHEETALA JYOTI (DVR-2)

This hybrid has been developed through heterosis breeding (PDIB-2 × PDIB-1) at IIVR, Varanasi and notified by CVRC Notification no. 1135(E) dated 15.11.2001. It is green fruited, resistant to YVMV as well as ELCV and has yield potential of 220-240 q/ha.

AZAD BHINDI-1

This variety has been developed at CSAUA&T, Kanpur and notified by CVRC Notification no. 1135(E) dated 15.11.2001. It is medium tall, erect, fruit medium long, thin, dark green with 5 ridges. It has yield potential of 275-300 q/ha.

GUJRAT OKRA-2

This variety has been developed at AAU, Anand and notified by CVRC Notification no. 1135(E) dated 15.11.2001. Its fruits are smooth, tender, and 16.7 cm long green pods with five ridges. It has yield potential of 150-170 q/ha.

KASHI PRAGATI (VRO-6)

This variety has been developed at IIVR, Varanasi and notified by CVRC Notification no. 597(E) dated 25.04.2006. It is tall (130-175 cm) with 2-3 slightly curved branches. First flower appears after 36-38 days after sowing on 4th nods during rainy season and 3rd node during summer season. Single plant bears 23-25 fruits of 8-10 cm length. It is resistant to YVMV and OLCV. It has been recommended for cultivation in Punjab, U.P., Bihar, Jharkhand, Chhatisgarh, Orissa and A.P. and has yield potential of 190-210 q/ha (rainy season crop) and 130-150 q/ha (summer season crop).

KASHI VIBHUTI (VRO-5)

This variety has been developed at IIVR, Varanasi and notified by CVRC Notification no. 597(E) dated 25.04.2006. It is 50-70 cm tall, short inter-nodes with two to three branching. Flowering starts at 4th or 5th nodes after 38-40 days after sowing. Fruits have 5 ridges, single plant bears 18-22 fruits with 8-10 length cm. Plants are resistant to YVMV and OLCV. It has been recommended for cultivation in Rajasthan, Gujarat, Haryana and Delhi and has yield potential of 180-200 q/ha.

KASHI SAATDHARI (IIVR-10)

This variety has been developed through pure line selection at IIVR, Varanasi and notified by CVRC Notification no. 98(E) dated 08.02.1997. It is 90-120 cm tall having 2-3 branches and 5-7cm internodal length. Flowering starts from 3rd to 4th nodes at 43 days after sowing. Its fruits have seven ridges and 13-15 cm length. Single plant bears 18 fruits. It has yield potential of 140-160 q/ha in rainy season and 110-125 q/ha in summer season.

PHULE UTKARSHA (GK-IV-3-3-3)

This variety has been developed through pedigree selection from a cross (Parbhani Kranti × Varsha Uphar) at MPKV, Rahuri and notified by CVRC Notification no. 597(E) dated 25.04.2006. Its fruits are dark green, slender and attractive. It is highly tolerant to YVMV and has yield potential of 300-325 q/ha in 100-110 days of crop duration.

HBH-142

This hybrid has been developed through heterosis breeding at HAU, Hisar, and notified by CVRC Notification no. 597(E) dated 25.04.2006.

It is resistant to YVMV. It has been recommended for cultivation throughout the country and has yield potential of 120-140 q/ha.

KASHI BHAIRAV (DVR-3)

This hybrid has been developed through heterosis breeding at IIVR, Varanasi and notified by CVRC Notification no. 597(E) dated 25.04.2006. Plants are medium tall with 2-3 branches, fruit colour dark green, length 10-12 cm at marketable stage. It is resistant to YVMV and OLCV under field condition. It has been recommended for cultivation throughout the country and has yield potential of 160-180 q/ha.

ARKA ANAMIKA (IIHR-10)

This variety has been developed at IIHR Bangalore and notified by CVRC Notification no. 2035(E) dated 28.11.2006. It is derived from interspecific cross between *Abelmoschus esculentus* and a wild species *A. manihot* spp. *tetraphyllus*. Plants are medium tall of about 100 cm with short internode length and fewer branches. Purple pigmentations are present on the stem, petiole and lower surface of the basal leaf. Leaves are green, small and deeply lobed. Stem petiole and leaves are sparsely hairy. It is an early maturing and takes about 50 days for first flowering after sowing and about 55 days for the first picking. Its fruits are medium, green, rough, 5-ridged and start after 5-6th node onward. It is YVMV resistant and has yield potential of 115-130 q/ha.

ARKA ABHAYA (IIHR SEL-4)

This variety has been developed at IIHR Bangalore and notified by CVRC Notification no. 2035(E) dated 28.11.2006. It is erect with a height of 170 cm and branched. Fruits are 18 cm in length having 5-6 ridges, dark green in colour with good keeping quality. It is resistant to YVMV and has yield potential of 180-200 q/ha in 120-130 days of crop duration.

HISAR NAVEEN (HRB-107-4)

This variety has been developed at HAU, Hisar and notified by CVRC Notification no. 2035(E) and 28.11.2006. Its plant height is 100-120 cm bearing smooth pods with 5 ridges, fruit length 9-10 cm and diameter 1.4 cm. It is moderately tolerant to leaf hopper and resistant to YVMV and has yield potential of 100-125 q/ha.

KASHI LILA (IIVR-11)

This variety has been developed at IIVR, Varanasi and notified by CVRC Notification no. 2035(E) dated 28.11.2006. Its plant height is 110-130 cm bearing 13-15 cm long green fruits with five ridges. First flower appears at 30-34 days after seed sowing. It is suitable for rainy and summer season cultivation. It is resistant to YVMV. It has been recommended for cultivation in Rajasthan, Gujarat, Haryana, Delhi, Madhya Pradesh and Maharashtra and has yield potential of 150-170 q/ha.

SUSTHIRA (AE-286-1)

This variety has been developed at KAU, Vellannikkara and notified by CVRC Notification no. 597(E) dated 25.04.2006. Fruits green with five ridges, 12 cm in length, first fruit on 6th node. It is a high yielding, mosaic tolerant, and long duration variety suitable for cultivation in *kharif* season and has yield potential of 180-200 q/ha.

KASHI MAHIMA (DVR-4)

This hybrid has been developed through heterosis breeding at IIVR, Varanasi and identified for release through AICRP-VC in 2001. It is tall (130-170 cm) bearing 12-14 cm long green fruits. Flowering starts at 36-40 days after sowing at 4-5 nodes. It is resistant to YVMV and OLCV in field conditions. It has been recommended for cultivation in Punjab, U.P., Bihar, Jharkhand, Chhatisgarh, Orissa, A.P., Madhya Pradesh and Maharashtra and has yield potential of 230-250 q/ha.

KASHI MOHINI (VRO-3)

This variety has been developed through hybridization followed by pedigree selection at IIVR, Varanasi and identified for release through AICRP-VC in 2001. Plants are tall (110-140 cm) and resistant to ELCV. Fruits are 11.3-12.6 cm long, green with five ridges. First flower appears at 4/5 node during summer and 5/7 nodes during rainy season after 39-41 days of sowing. It is suitable for summer and rainy season. It stands at high temperature during summer and resistant to OLCV. It has been recommended for cultivation throughout the country and has yield potential of 130-150 q/ha.

KASHI MANGALI (VRO-4)

This variety has been developed through hybridization followed by pedigree selection at IIVR, Varanasi and identified for release through AICRP-VC in 2002. Plants are tall and resistant to ELCV. Its fruits are green with five ridges. Flowers appear on 4/5 node. It is resistant to ELCV and suitable for cultivation in summer and rainy season. It has been recommended for cultivation in Punjab, U.P., Bihar, Chhattisgarh, Orissa and Andhra Pradesh and has yield potential of 160-175 q/ha.

NDO-10

This variety has been developed at NDUA&T, Faizabad and identified for release through AICRP-VC in 2005. Its plants are 50-100 cm in height bearing 12-15 cm long dark green fruits with 5 ridges. Fruits are ready to harvest in 35-40 days in summer and 45-50 days in rainy season. It is resistant to YVMV and has yield potential of 50-55 q/ha (summer crop) and 75-80 q/ha in rainy season.

GUJRAT OKTA HYBRID-2 (JNDOH-02-2)

This hybrid has been developed through heterosis breeding (KS-404 × HRB-108-2) at JAU, Junagadh and identified for release through AICRP-VC in 2008. It is tall (130-170 cm) bearing 12-14 cm long and tender green pods. Flowering starts at 36-40 days after sowing at 4-5 nodes. It is suitable for *kharif* season sowing. It has been recommended for cultivation in West Bengal, Assam, Uttaranchal, Orissa, Andhra Pradesh, Rajsthan, Gujrat, Haryana, Delhi, M.P., Maharashtra, Karnataka, Kerala and Tamilnadu and has yield potential of 140-150 q/ha.

Notified Varieties/Hybrids of Vegetable Crops

Crop/ Variety	Deveoping Centre	Year	DAC Notification Number[§] and Date
	SOLANACEOUS VEGETABLES		
Tomato			
Sioux	IARI, Katrain	1970	716, 20.02.1970
Sioux	IARI, Katrain	1973	361(E), 30.06.1973
Sioux	IARI, Katrain	1975	440, 21.08.1975
Pusa Ruby	IARI, New Delhi	1970	716, 20.02.1970
Pusa Ruby	IARI, New Delhi	1973	361(E), 30.06.1973
Pusa Ruby	IARI, New Delhi	1975	440, 21.08.1975
Pusa Ruby	IARI, New Delhi	1985	540(E), 24.07.1985
S-12	PAU, Ludhiana	1970	716, 20.02.1970
S-12	PAU, Ludhiana	1978	1004, 23.03.1978
Marglobe	IARI, New Delhi	1970	716, 20.02.1970
Marglobe	IARI, New Delhi	1973	361(E), 30.06.1973
Marglobe	IARI, New Delhi	1975	440, 21.08.1975
Marglobe	IARI, New Delhi	1978	13, 19.12.1978
SL-120	IARI, New Delhi	1970	716, 20.02.197
Best of All	IARI, New Delhi	1973	361(E), 30.06.1973
Best of All	IARI, New Delhi	1975	440, 21.08.1975
Best of All	IARI, New Delhi	1978	13, 19.12.1978
Pusa Early Dwarf	NBPGR, New Delhi	1973	361(E), 30.06.1973

Crop/ Variety	Deveoping Centre	Year	DAC Notification Number[§] and Date
Pusa Early Dwarf	NBPGR, New Delhi	1975	440, 21.08.1975
Pusa Early Dwarf	NBPGR, New Delhi	1985	540(E), 24.07.1985
Punjab Selection No.-12	PAU, Ludhiana	1975	440, 21.08.1975
Solan Gola	HPKV, Solan	1975	440, 21.08.1975
HS-101	HAU, Hisar	1976	786, 02.02.1976
HS-101	HAU, Hisar	1978	1004, 23.03.1978
HS-102	HAU, Hisar	1976	786, 02.02.1976
CO-2	TNAU, Coimbatore	1978	13, 19.12.1978
Keckruth*	Introduction	1978	13, 19.12.1978
Keckruth*	Introduction	1991	563(E), 30.08.1991
Keckruth Ageti*	Introduction	1978	13, 19.12.1978
Keckruth Ageti*	Introduction	1991	563(E), 30.08.1991
Punjab Chhuhara	PAU, Ludhiana	1978	13, 19.12.1978
Punjab Tropic	PAU, Ludhiana	1978	13, 19.12.1978
NTDR-1	NDUAT, Faizabad	1982	19(E), 14.01.1982
Kaliyanpur Angoorlata	CSAUAT, Kanpur	1983	2(E), 03.01.1983
Kaliyanpur T-1 (T-1)	CSAUAT, Kanpur	1983	2(E), 03.01.1983
Junagarh Ruby	GAU, Junagarh	1984	596(E), 13.08.1984
Marutham (CO-3)	TNAU, Coimbatore	1984	596(E), 13.08.1984
Azad T-2	CSAUAT, Kanpur	1985	295(E), 09.04.1985
Pant Bahar	GBPUAT, Pantnagar	1985	540(E), 24.07.1985
PKM-1	TNAU, Coimbatore	1985	540(E), 24.07.1985

Crop/ Variety	Deveoping Centre	Year	DAC Notification Number[§] and Date
Punjab NR-7	PAU, Ludhiana	1987	165(E), 06.03.1987
Pusa Gaurav	IARI, New Delhi	1988	10(E), 01.01.1988
Pant T-3	GBPUAT, Pantnagar	1988	1135(E), 01.12.1988
ATV-1	MPKV, Rahuri	1989	915(E), 06.11.1989
ATH-1 (Hybrid)	MPKV, Rahuri	1989	915(E), 06.11.1989
Sonali (Sel-2)	–	1989	915(E), 06.11.1989
Arka Vikas (Selection-22)	IIHR, Bangalore	1990	386(E), 15.05.1999
Utkal Pallavi	OUAT, Bhubaneshwar	1992	860(E), 25.11.1992
Hisar Lalit (NT-8)	HAU, Hisar	1993	617(E), 17.08.1993
Hisar Arun (Sel-7)	HAU, Hisar	1994	636(E), 02.09.1994
Narendra Tomato-2	NDUAT, Faizabad	1996	115(E), 10.02.1996
Pusa Hybrid-2	IARI, New Delhi	1996	115(E), 10.02.1996
Rajashri Hybrid-2	UAS, Dharwar	1996	115(E), 10.02.1996
Solan Sagun	YSPUHF, Solan	1996	115(E), 10.02.1996
Pusa Divya (KT-4)	IARI, New Delhi	1997	98(E), 08.02.1997
Pusa Hybrid-4 (DTH-4)	IARI, New Delhi	1997	98(E), 08.02.1997
Utkal Kumari (BT-10)	OUAT, Bhubaneshwar	1998	843(E), 21.09.1998
Solan Vajr	YSUHF, Solan	1998	843(E), 21.09.1998
TH-802	IIHR, Bangalore	1998	843(E), 21.09.1998
Mukthi (L-79-5)	KAU, Vellanikkara	1998	843(E), 21.09.1998
Vasundhara (Hybrid-28)	MPKV, Rahuri	2001	1135(E), 15.11.2001

Crop/ Variety	Deveoping Centre	Year	DAC Notification Number[§] and Date
Parbhani Yashshri (Sel-14)	MPKV, Rahuri	2001	1135(E), 15.11.2001
Swarna Baibhav (CHTH-1)	HARP, Ranchi	2004	642 (E), 31.05.2004
Sankranti (TLB-11)	IIHR, Bangalore	2004	642 (E), 31.05.2004
Nandi (TLB-130)	IIHR, Bangalore	2004	642(E), 31.05.2004
Vaibhav (TLB-182)	IIHR, Bangalore	2004	642(E), 31.05.2004
TH-1	IIHR, Bangalore	2005	664(E), 11.05.2005
Kashi Vishesh (H-86)	IIVR, Varanasi	2006	597(E), 25.04.2006
Kashi Amrit (DVRT-1)	IIVR, Varanasi	2006	597(E), 25.04.2006
Kashi Anupam (DVRT-2)	IIVR, Varanasi	2006	597(E), 25.04.2006
Kashi Sharad (IIVR Sel-1)	IIVR, Varanasi	2006	597(E), 25.04.2006
Kashi Hemant (IIVR Sel-2)	IIVR, Varanasi	2006	597(E), 25.04.2006
Arka Ananya (TLBRH-9)	IIHR, Bangalore	2006	597(E), 25.04.2006
Arka Meghali	IIHR, Bangalore	2006	2035(E), 28.11.2006
Swarna Naveen	HARP, Ranchi	2006	2035(E), 28.11.2006
Swarna Lalima	HARP, Ranchi	2006	2035(E), 28.11.2006
Swarna Sampada (HATH-3)	HARP, Ranchi	2006	2035(E), 28.11.2006
Mani Leima	ICARRC, Barapani	2006	2035(E), 28.11.2006
Azad T-6	CSAUAT, Kanpur	2006	597(E), 25.04.2006

Crop/ Variety	Deveoping Centre	Year	DAC Notification Number[§] and Date
Narendra Tomato-5 (NDT-96)	NDUAT, Faizabad	2006	597(E), 25.04.2006
Narendra Tomato-6 (NDT-4)	NDUAT, Faizabad	2006	597(E), 25.04.2006
GT-2	GAU, Anand	2006	597(E), 25.04.2006
Brinjal			
Pusa Purple Long	IARI, New Delhi	1970	716, 20.02.1970
Pusa Purple Long	IARI, New Delhi	1973	361(E), 30.06.1973
Pusa Purple Long	IARI, New Delhi	1975	440, 21.08.1975
Pusa Purple Long	IARI, New Delhi	1976	786, 02.02.1976
Pusa Purple Long	IARI, New Delhi	1985	540(E), 20.07.1985
Pusa Purple Round	IARI, New Delhi	1970	716, 20.02.1970
Pusa Purple Round	IARI, New Delhi	1975	440, 21.08.1975
Pusa Purple Round	IARI, New Delhi	1976	786, 02.02.1976
Pusa Purple Round	IARI, New Delhi	1978	13, 19.12.1978
Pusa Purple Round	IARI, New Delhi	1985	540(E), 20.07.1985
Punjab No.-8	PAU, Ludhiana	1973	361(E), 30.06.1973
R-34	PAU, Ludhiana	1973	361(E), 30.06.1973
Black Beauty	Introduction	1975	440, 21.08.1975
Black Beauty	Introduction	1978	13, 19.12.1978
Pusa Cluster	IARI, New Delhi	1975	440(E), 21.08.1975
Pusa Purple Cluster	IARI, New Delhi	1978	1004, 23.03.1978
Brinjal (Kind)	-	1975	441(E), 21.08.1975
H-4	HAU, Hisar	1976	786, 02.02.1976

Crop/ Variety	Deveoping Centre	Year	DAC Notification Number[§] and Date
Punjab Haryana Brinjal-4	PAU, Ludhiana	1976	1004, 23.03.1978
Jamuni Gol Baingan	PAU, Ludhiana	1978	1004, 23.03.1978
Eranagere	TNAU, Coimbatore	1978	13, 19.12.1978
Punjab Bahar	PAU, Ludhiana	1978	13, 19.12.1978
Punjab Chamkila	PAU, Ludhiana	1978	13, 19.12.1978
Bhagyamati	ANGRAU, Hyderabad	1982	19(E), 14.01.1982
MDU-1	TNAU, Coimbatore	1982	19(E), 14.01.1982
Kalyanpur T-3 (T-3)	CSAUAT, Kanpur	1983	2(E), 03.01.1983
Junagarh Long	GAU, Junagarh	1984	596(E), 13.08.1984
Junagarh Oblong	GAU, Junagarh	1984	596(E), 13.08.1984
Gujrat-6	GAU, Anand	1984	596(E), 13.08.1984
Azad Kranti	CSAUAT, Kanpur	1985	295(E), 09.04.1985
Azad B-1	CSAUAT, Kanpur	1985	295(E), 09.04.1985
Pant Samrat	GBPUAT, Pantnagar	1985	295(E), 09.04.1985
Pant Rituraj	GBPUAT, Pantnagar	1985	540(E), 24.07.1985
CO-1	TNAU, Coimbatore	1985	540(E), 24.07.1985
Annamalai	TNAU, Coimbatore	1985	832(E), 18.11.1985
Arka Shirish	IIHR, Bangalore	1986	867(E), 26.11.1986
Arka Kusumakar	IIHR, Bangalore	1986	867(E), 26.11.1986
Arka Sheel	IIHR, Bangalore	1986	867(E), 26.11.1986
Arka Navneet	IIHR, Bangalore	1986	867(E), 26.11.1986
Jamuni Gola (Sel-16)	PAU, Ludhiana	1987	165(E), 06.03.1987

Crop/ Variety	Deveoping Centre	Year	DAC Notification Number[§] and Date
Punjab Barsati	PAU, Ludhiana	1989	280(E), 13.04.1989
Sada Bahar Baigan	PAU, Ludhiana	1989	280(E), 13.04.1989
Aruna	PDKV, Akola	1989	915(E), 06.11.1989
Pusa Anupam	IARI, New Delhi	1991	527(E), 16.08.1991
Utkal Tarini (BB-7)	OUAT, Bhubaneshwar	1992	860(E), 25.11.1992
Pusa Hybrid-6	IARI, New Delhi	1993	617(E), 17.08.1993
Hisar Shyamal (H-8)	HAU, Hisar	1993	617(E), 17.08.1993
Hisar Pragati (H-7)	HAU, Hisar	1994	636(E), 02.09.1994
Pusa Hybrid-5	IARI, New Delhi	1994	636(E), 02.09.1994
Pant Brinjal Hybrid-1	GBPUAT, Pantnagar	1994	636(E), 02.09.1994
BH-1	OUAT, Bhubaneshwar	1996	115(E), 10.02.1996
BH-2	OUAT, Bhubaneshwar	1996	115(E), 10.02.1996
GBH-1	GAU, Anand	1996	115(E), 10.02.1996
Narendra Baigan-1	NDUAT, Faizabad	1996	115(E), 10.02.1996
Narendra Hybrid Brinjal-1 (NDBH-1)	NDUAT, Faizabad	1996	115(E), 10.02.1996
Pusa Uttam (DBR-31)	IARI, New Delhi	1997	98(E), 08.02.1997
Pusa Upkar (DBR-8)	IARI, New Delhi	1997	98(E), 08.02.1997
Pusa Bindu (DBSR-44)	IARI, New Delhi	1997	98(E), 08.02.1997
Utkal Madhuri (BB-44)	OUAT, Bhubaneshwar	1998	834(E), 21.01.1998

Crop/ Variety	Deveoping Centre	Year	DAC Notification Number[§] and Date
Utkal Keshari (BB-26)	OUAT, Bhubaneshwar	1998	834(E), 21.01.1998
JC-1	IGKV, Jabalpur	1998	834(E), 21.01.1998
JC-2	IGKV, Jabalpur	1998	834(E), 21.01.1998
Sweta (SM-6-6)	KAU, Vellanikkara	1998	834(E), 21.01.1998
Pusa Ankur (DBSR-91)	IARI, New Delhi	1999	1052(E), 26.10.1999
Azad Hybrid-1	CSAUAT, Pantnagar	1999	1052(E), 26.10.1999
KKM-1 (KSM-107)	TNAU, Coimbatore	1999	1052(E), 26.10.1999
PLR-1	TNAU, Coimbatore	2001	1135(E), 15.11.2001
Gujarat Brinjal Long-1 (GBL-1)	GAU, Anand	2003	448(E), 21.04.2003
Brinjal Hybrid-2 (GBH-2)	GAU, Anand	2004	642(E), 31.05.2004
Swarna Mani (CHBR-1)	HARP, Ranchi	2004	642(E), 31.05.2004
Swarna Pratibha (CH-309)	HARP, Ranchi	2004	642(E), 31.05.2004
Utkal Jyoti (BB-13)	OUAT, Bhubaneswar	2004	642(E), 31.05.2004
Kashi Prakash (IVBR-1)	IIVR, Varanasi	2006	597(E), 25.04.2006
Kashi Taru (IVBL-9)	IIVR, Varanasi	2006	597(E), 25.04.2006
Swarna Shobha (HABR-4)	HARP, Ranchi	2006	597(E), 25.04.2006
Narendra Hybrid Brinjal-3 (NDBH-3)	NDUAT, Faizabad	2006	597(E), 25.04.2006

Crop/ Variety	Deveoping Centre	Year	DAC Notification Number[§] and Date
Gujarat Oblong Brinjal-2	GAU, Anand	2006	597(E), 25.04.2006
Phule Harit (RHRB-16)	MPKV, Rahuri	2006	2035(E), 28.11.2006
Swarna Shyamali	HARP, Ranchi	2006	2035(E), 28.11.2006
Swarna Shree	HARP, Ranchi	2006	2035(E), 28.11.2006
Swarna Ajay (HABH-3)	HARP, Ranchi	2006	2035(E), 28.11.2006
Kashi Sandesh (VRBHR-1)	IIVR, Varanasi	2006	2035(E), 28.11.2006
Arka Keshav (IIHR-21)	IIHR, Bangalore	2006	2035(E), 28.11.2006
Arka Nidhi (IIHR-12)	IIHR, Bangalore	2006	2035(E), 28.11.2006
Chilli			
Hot Pourtgal	RARS, Lam	1975	440, 21.08.1975
Sanauri	RARS, Lam	1975	440, 21.08.1975
Sweet Banana	RARS, Lam	1975	440, 21.08.1975
Chilli (Kind)	–	1975	440, 21.08.1975
Andhra Jyoti (G-5)	RARS, Lam	1978	1004, 23.03.1978
Sindhur (CA-960)	RARS, Lam	1978	13, 19.12.1978
Pant C-1	GBPUAT, Pantnagar	1982	19(E), 14.01.1982
CO-1	TNAU, Coimbatore	1982	19(E), 14.01.1982
DH-76-6	UAS, Dharwad	1982	19(E), 14.01.1982
Jwala (Pusa Jwala)	IARI, New Delhi	1983	499(E), 08.07.1983
CO-2	TNAU, Coimbatore	1984	596(E), 13.08.1984

Crop/ Variety	Deveoping Centre	Year	DAC Notification Number[§] and Date
Chanchal	CSAUAT, Kanpur	1985	295(E), 09.04.1985
K-2	TNAU, RS, Kovilpatti	1985	540(E), 24.07.1985
MDU-1	TNAU, Coimbatore	1985	540(E), 24.07.1985
K-1	TNAU, RS, Kovilpatti	1985	832(E), 18.11.1985
Bhaskar (X-235; LCA- 235)	RARS, Lam	1986	867(E), 26.11.1986
Punjab Lal	PAU, Ludhiana	1987	165(E), 06.03.1987
Jawahar Mirch-218 (J-218)	JNKVV, Jabalpur	1989	915(E), 06.11.1989
Musalwadi	MPKV, Rahuri	1991	527(E), 16.08.1991
PKM-1	TNAU, Coimbatore	1991	793(E), 22.11.1991
Konkan Kirti	KKV, Konkan	1993	617(E), 17.08.1993
Lam-305	RARS, Lam	1995	408(E), 04.05.1995
Jayanti (AKC-86-39)	PDKV, Akola	1996	115(E), 10.02.1996
Parbhani Tejas	MAU, Parbhani	1996	115(E), 10.02.1996
Punjab Guchhedar	PAU, Ludhiana	1996	115(E), 10.02.1996
Punjab Surkh (ELS-2)	PAU, Ludhiana	1996	115(E), 10.02.1996
Gaura Chilli-2	–	1999	1052(E), 26.10.1999
PLR-1	TNAU, Coimbatore	1999	1052(E), 26.10.1999
Jawahar Mirch-283	JNKVV, Jabalpur	1999	1052(E), 26.10.1999
Kashmir Long-1	SKUAT, Srinagar	2001	1135(E), 15.11.2001
PMK-1	TNAU, Coimbatore	2001	1135(E), 15.11.2001
CO-3	TNAU, Coimbatore	2001	1135(E), 15.11.2001

Crop/ Variety	Deveoping Centre	Year	DAC Notification Number[§] and Date
Utkal Ava (BC 14-2)	OUAT, Bhubaneswar	2004	642(E), 31.05.2004
Arka Suphal (PMR 57/88 K)	IIHR, Bangalore	2006	597(E), 25.04.2006
Gujarat Veg. Chilli (GVC-101)	AAU, Anand	2006	597(E), 25.04.2006
Kashi Anmol (KA-2)	IIVR, Varanasi	2006	597(E), 25.04.2006
Azad Mirch-1	CSAUAT, Kanpur	2006	597(E), 25.04.2006
Anugraha	KAU, Vellanikkara	2006	597(E), 25.04.2006
Gujarat Veg. Chilli-111 (GVC-111)	GAU, Anand	2006	597(E), 25.04.2006
Gujarat Veg. Chilli-121 (ACS-97-2)	GAU, Anand	2006	597(E), 25.04.2006
Prasanth (LCA-334)	RARS, Lam	2006	597(E), 25.04.2006
Arka Lohit	IIHR, Bangalore	2006	2035(E), 28.11.2006
Arka Meghna (MSH-172)	IIHR, Bangalore	2006	2035(E), 28.11.2006
Kashi Surkh (CCH-2)	IIVR, Varanasi	2006	2035(E), 28.11.2006
RCH-1	ARS, Durgapura	2004	642(E), 31.05.2004
Sweet Pepper			
California Wonder	IARI-RS, Katrain	1975	440, 21.08.1975
California Wonder	IARI-RS, Katrain	1978	13, 19.12.1978
Chinese Giant	IARI-RS, Katrain	1975	440, 21.08.1975
Chinese Giant	IARI-RS, Katrain	1978	13, 19.12.1978

Crop/ Variety	Deveoping Centre	Year	DAC Notification Number§ and Date
Hungarian	IARI-RS, Katrain	1975	440, 21.08.1975
Bull Nose	IARI-RS, Katrain	1978	13, 19.12.1978
Arka Basant	IIHR, Bangalore	1986	867(E), 26.11.1986
Arka Gaurav	IIHR, Bangalore	1986	867(E), 26.11.1986
Arka Mohini	IIHR, Bangalore	1986	867(E), 26.11.1986
Solan Hybrid-1	YSPUHF, Solan	1996	115(E), 10.02.1996
Pusa Deepti (KT-1)	IARI, Katrain	1997	98(E), 08.02.1997
Nishat-1 (Capsicum Sel-2)	SKUAST, Srinagar	2001	1135(E), 15.11.2001
	LEGUME VEGETABLES		
Vegetable Pea			
Bonneville	IARI, New Delhi	1970	716, 20.02.1970
Bonneville	IARI, New Delhi	1973	361(E), 30.06.1973
Bonneville	IARI, New Delhi	1975	440, 21.08.1975
Arkel	IARI, New Delhi	1975	1004, 23.03.1978
VL-3	VPKAS, Almora	1988	1135(E), 01.12.1988
Pant Matar-2 (PM-2)	GBPUAT, Pantnagar	1989	915(E), 06.11.1989
VL-7	VPKAS, Almora	1995	408(E), 04.05.1995
Ageta-6	PAU, Ludhiana	1996	115(E), 10.02.1996
Mithi Phalli (Oregon Sugar Pod)	PAU, Ludhiana	1996	115(E), 10.02.1996
VL-6	VPKAS, Almora	1997	98(E), 08.02.1997
NDVP-8	NDUAT, Faizabad	1998	843(E), 21.09.1998

Crop/ Variety	Deveoping Centre	Year	DAC Notification Number[§] and Date
Vivek Matar-8 (VL-8)	VPKAS, Almora	2001	1135(E), 15.11.2001
Vivek Matar-9	VPKAS, Almora	2005	664(E), 11.05.2005
Kashi Nandini (VRP-5)	IIVR, Varanasi	2006	597(E), 25.04.2006
Narendra Sabzi Matar-6 (NDVP-12)	NDUAT, Faizabad	2006	597(E), 25.04.2006
Narendra Sabzi Matar-4 (NDVP-9)	NDUAT, Faizabad	2006	597(E), 25.04.2006
Narendra Sabzi Matar-5 (NDVP-250)	NDUAT, Faizabad	2006	597(E), 25.04.2006
Kashi Uday (VR-6)	IIVR, Varanasi	2006	597(E), 25.04.2006
Kashi Shakti (VR-7)	IIVR, Varanasi	2006	597(E), 25.04.2006
Kashi Mukti (VR-22)	IIVR, Varanasi	2006	597(E), 25.04.2006
Azad P-5 (KS-245)	CSAUAT, Kanpur	2006	2035(E), 28.11.2006
Swarna Mukti (CHP-2)	HARP, Ranchi	2006	2035(E), 28.11.2006
Arka Ajit (FC-1)	IIHR, Bangalore	2006	2035(E), 28.11.2006
Cowpea			
Pusa Barsati	IARI, New Delhi	1974	566(E), 21.09.1974
Pusa Phaguni	IARI, New Delhi	1974	566(E), 21.09.1974
Pusa Rituraj (Pusa-152)	IARI, New Delhi	1978	1004, 23.03.1978
S-488	IARI, New Delhi	1980	2103, 12.08.1980
Birsa Sweta	BAU, Ranchi	1986	258(E), 14.05.1986

Crop/ Variety	Deveoping Centre	Year	DAC Notification Number[§] and Date
Pusa Komal (Sel-1552)	IARI, New Delhi	1987	165(E), 06.03.1987
Cowpea-263 (Sel-263)	IIHR, Bangalore	1994	636(E), 02.09.1994
Narendra Lobia-1	NDUAT, Faizabad	1996	115(E), 10.02.1996
Kashi Shyamal (IVRCP-1)	IIVR, Varanasi	2006	597(E), 25.04.2006
Kashi Gauri (IVRCP-2)	IIVR, Varanasi	2006	597(E), 25.04.2006
Anaswara	KAU, Vellanikkara	2006	597 (E), 25.04.2006
Arka Suman	IIHR, Bangalore	2006	2035(E), 28.11.2006
Arka Samrudhi (IIHR-16)	IIHR, Bangalore	2006	2035(E), 28.11.2006
Swarna Suphala (CHCP-2)	HARP, Ranchi	2006	2035(E), 28.11.2006
French Bean			
Premier	NBPGR, New Delhi	1975	440, 21.08.1975
Kentucky Wonder	NBPGR, New Delhi	1975	440, 21.08.1975
Contender	NBPGR, New Delhi	1975	440, 21.08.1975
VL-1	VPKAS, Almora	1978	13, 19.12.1978
VL Boni-1	VPKAS, Almora	1978	13, 19.12.1978
VL-2	VPKAS, Almora	1978	13, 19.12.1978
VL Rajmah-63	VPKAS, Almora	1982	19(E), 14.01.1982
Arka Komal (Sel-9)	IIHR, Bangalore	1990	386(E), 15.05.1990
Pant Anupama (UPF-191)	GBPUAT, Pantnagar	1996	115(E), 10.02.1996

Crop/ Variety	Deveoping Centre	Year	DAC Notification Number[§] and Date
Azad Rajmah-1	CSAUAT, Kanpur	2006	597(E), 25.04.2006
Phule Suyash (GK-7)	MPKV, Rahuri	2006	597(E), 25.04.2006
Kashi Param (IVFB-1)	IIVR, Varanasi	2006	2035(E), 28.11.2006
Arka Suvidha (IIHR-909)	IIHR, Bangalore	2006	2035(E), 28.11.2006
Indian Bean			
Hebbal Awarai-3	IIHR, Bangalore	1982	19(E), 14.01.1982
CO-1	TNAU, Coimbatore	1982	19(E), 14.01.1982
Wal Konkan-1	PDKV, Dapoli	1985	295(E), 09.04.1985
CO-2	TNAU, Coimbatore	1985	540(E), 24.07.1985
Konkan Bhushan	KKV, Dapoli	1993	617(E), 17.08.1993
Phule Gauri (RHRWL-1)	MPKV, Rahuri	2006	2035(E), 28.11.2006
Swarna Utkrisht (CHDB-1)	HARP, Ranchi	2006	2035(E), 28.11.2006
BULB VEGETABLES			
Onion			
Punjab Selection	PAU, Ludhiana	1973	361(E), 30.06.1973
Onion (Kind)	–	1975	441(E), 21.08.1975
Hisar-2	HAU, Hisar	1976	786, 02.02.1976
CO-2	TNAU, Coimbatore	1978	13, 19.12.1978
Punjab-48 (S-48)	PAU, Ludhiana	1978	13, 19.12.1978

Crop/ Variety	Deveoping Centre	Year	DAC Notification Number§ and Date
Pusa Ratnar	IARI, New Delhi	1978	13, 19.12.1978
Pusa Red	IARI, New Delhi	1978	13, 19.12.1978
VL Piaza-67	VPKAS, Almora	1978	13, 19.12.1978
CO-3	TNAU, Coimbatore	1982	19(E), 14.01.1982
MDU-1	MPKV, Rahuri	1982	19(E), 14.01.1982
Kaliyanpur Red Round	CSAUAT, Kanpur	1983	2(E), 03.01.1983
CO-4	TNAU, Coimbatore	1984	596(E), 13.08.1984
Agri Found Dark Red	NHRDF, Nasik	1988	1135(E), 01.12.1998
Baswant (No.-780)	IIHR, Bangalore	1989	915(E), 06.11.1989
Pusa Madhavi (Line-120)	IARI, New Delhi	1989	915(E), 06.11.1989
Arka Niketan (Sel-13)	IIHR, Bangalore	1990	386(E), 15.05.1999
Arka Kalyan (Sel-14)	IIHR, Bangalore	1990	386(E), 15.05.1999
VL Piaza-3	VPKAS, Almora	1991	793(E), 22.11.1991
Agri Found Light Red	NHRDF, Nasik	1996	115(E), 10.02.1996
Punjab Naroya (PBR-5)	PAU, Ludhiana	1996	115(E), 10.02.1996
Punjab White	PAU, Ludhiana	1998	843(E), 21.09.1998
Phule Suvarna (RHR-87015)	MPKV, Rahuri	2001	348(E), 19.04.2001
Rajasthan Onion-1 (RO-1)	RARS, Durgapura	2004	642(E), 31.05.2004
Arpita (RO-59)	RARS, Durgapura	2005	664(E), 11.05.2005

Crop/ Variety	Deveoping Centre	Year	DAC Notification Number§ and Date
Arka Pitamber	IIHR, Bangalore	2006	2035(E), 28.11.2006
Arka Bindu	IIHR, Bangalore	2006	2035(E), 28.11.2006
NHRDF Red (L-28)	NHRDF, Karnal	2006	2035(E), 28.11.2006
Akola Safed (PKV Sel White)	PDKV, Akola	2006	597(E), 25.04.2006
Phule Samrath (S-1)	MPKV, Rahuri	2006	597(E), 25.04.2006
Garlic			
Agrifound White (G-41)	NHRDF, Karnal	1989	280(E), 13.04.1989
Yamuna Safed (G-1)	NHRDF, Karnal	1991	527(E), 16.08.1991
Godawari (Sel-2)	NHRDF, Karnal	1993	617(E), 17.08.1993
Yamuna Safed-2 (G-50)	NHRDF, Karnal	1996	115(E), 10.02.1996
GHC-1	HPKV, Palampur	1996	115(E), 10.02.1996
Yamuna Safed-3	NHRDF, Karnal	1999	1052(E), 26.10.1999
VL Garlic-1 (VLG-7)	VPKAS, Almora	2004	642(E), 31.05.2004,
Yamuna Safed-4 (G-323)	NHRDF, Karnal	2006	597(E), 25.04.2006
COLE VEGETABLES			
Cauliflower			
Pusa Snowball-16	IARI, RS, Katrain	1970	716, 20.02.1970
Pusa Snowball-16	IARI, RS, Katrain	1973	361(E), 30.06.1973

Crop/ Variety	Deveoping Centre	Year	DAC Notification Number§ and Date
Pusa Snowball-16	IARI, RS, Katrain	1975	440, 21.08.1975
Pusa Snowball-16	IARI, RS, Katrain	1978	13, 19.12.1978
Pusa Katki-7	IARI, New Delhi	1970	716, 20.02.1970
Early Kuwari	PAU, Ludhiana	1973	361(E), 30.06.1973
Early Kuwari	PAU, Ludhiana	1978	1004, 23.03.1970
Giant Snowball	IARI, RS, Katrain	1973	361(E), 30.06.1973
Early Bhagot	IARI, RS, Katrain	1975	440, 21.08.1975
Pusa Katki	IARI, New Delhi	1975	440, 21.08.1975
D-96	IARI, RS, Katrain	1975	440, 21.08.1975
Hisar-1	HAU, Hisar	1976	786, 02.02.1976
Pusa Katki	IARI, New Delhi	1976	786, 02.02.1976
Pusa Katki	IARI, New Delhi	1978	13, 19.12.1978
Pusa Deepali	IARI, New Delhi	1978	1004, 23.03.1978
Pusa Snowball-1	IARI, RS, Katrain	1978	1004, 23.03.1978
Pusa Snowball-2	IARI, RS, Katrain	1978	1004, 23.03.1978
Pusa Shubhra	IARI, New Delhi	1985	295(E), 09.04.1985
Pusa Shubhra	IARI, New Delhi	1988	10(E), 01.01.1988
Pusa Snowball K-1	IARI, RS, Katrain	1988	10(E), 01.01.1988
Pant Gobhi-3 (235-5)	GBPUAT, Pantnagar	1989	280(E), 13.04.1989
Narendra Gobhi-1	NDUAT, Faizabad	1992	814(E), 04.11.1992
Pusa Early Synthetic	IARI, New Delhi	1993	617(E), 17.08.1993
Pusa Hybrid-2	IARI, New Delhi	1994	636(E), 02.09.1994

Crop/ Variety	Deveoping Centre	Year	DAC Notification Number[§] and Date
Pant Gobhi-4 (235-S)	GBPUAT, Pantnagar	1995	408(E), 04.05.1995
Pusa Sharad (Sel-309-1-2)	IARI, New Delhi	2004	642(E), 31.05.2004
Pusa Meghna (DC/98-2)	IARI, New Delhi	2004	642(E), 31.05.2004
Kashi Kuwanri (IVCE-2)	IIVR, Varanasi	2006	597(E), 25.04.2006
RCJOB-1	Jobner, Rajasthan	2006	597(E), 25.04.2006
Cabbage			
Golden Acre	IARI, RS, Katrain	1970	716, 20.02.1970
Golden Acre	IARI, RS, Katrain	1973	361(E), 30.06.1973
Golden Acre	IARI, RS, Katrain	1975	440, 21.08.1975
Golden Acre	IARI, RS, Katrain	1976	786, 02.02.1976
Golden Acre	IARI, RS, Katrain	1978	13, 19.12.1978
Golden Acre	IARI, RS, Katrain	1985	540(E), 24.07.1985
Pusa Drumhead	IARI, RS, Katrain	1970	716, 20.02.1970
Pusa Drumhead	IARI, RS, Katrain	1985	540(E), 24.07.1985
Copenhagen	IARI, RS, Katrain	1970	716, 20.02.1970
Pride of India	YSPUHF, Solan	1973	361(E), 30.06.1973
Pride of India	YSPUHF, Solan	1975	440, 21.08.1975
Pride of India	YSPUHF, Solan	1985	540(E), 24.07.1985
Cabbage (Kind)	IARI, RS, Katrain	1975	441, 21.08.1975
Early Drum Head	IARI, RS, Katrain	1978	13, 19.12.1978

Crop/ Variety	Deveoping Centre	Year	DAC Notification Number[§] and Date
Copenhagen (Market)	IARI, RS, Katrain	1978	13, 19.12.1978
Pusa Mukta	IARI, RS, Katrain	1989	280(E), 13.04.1988
Chini Sason	HPKV, Palampur	1989	10(E), 01.01.1989
PUCH-3 (Palampur Green)	HPKV, Palampur	1996	115(E), 10.02.1996
Kinner Red (Red Cabbage)	HPKV, Palampur	1999	1052(E), 26.10.1999
Broccoli			
Palam Haritica	HPKV, Palampur	2005	664(E), 11.05.2005
Palam Vachitra	HPKV, Palampur	2005	664(E), 11.05.2005
Palam Kanchan	HPKV, Palampur	2005	664(E), 11.05.2005
Knol Khol			
Large Green	IARI, RS, Katrain	1975	440, 21.08.1975
White Vienna	IARI, RS, Katrain	1975	440, 21.08.1975
Knol Khol (Kind)	IARI, RS, Katrain	1975	441(E), 21.08.1975
Early Purple Vienna	IARI, RS, Katrain	1978	13, 19.12.1978
King of Market	IARI, RS, Katrain	1978	13, 19.12.1978
Early White Vienna	IARI, RS, Katrain	1978	13, 19.12.1978
Palam Tender Knob	HPKV, Palampur	2005	664(E), 11.05.2005
CUCURBITACEOUS VEGETABLES			
Bitter Gourd			
Kashmiri Long	SKUAT, Srinagar	1975	440, 21.08.1975
Long yellow (Coimbatore Long)	TNAU, Coimbatore	1975	440, 21.08.1975

Crop/ Variety	Deveoping Centre	Year	DAC Notification Number§ and Date
Bitter Gourd (Kind)	–	1975	441(E), 21.08.1975
CO-1	TNAU, Coimbatore	1980	2103, 12.08.1980
Kaliyanpur Barahamasi	CSAUAT, Kanpur	1983	2(E), 03.01.1983
MDU-1	MPKV, Rahuri	1985	540(E), 24.07.1985
Arka Harit	IIHR, Bangalore	1986	867(E), 26.11.1986
Punjab-14	PAU, Ludhiana	1987	165(E), 06.03.1987
Kalyanpur Sona	CSAUAT, Kanpur	1996	115, 10.02.1996
Konkan Tara (DPL-BG-2)	KKV, Konkan	1996	115, 10.02.1996
Preethi	KAU, Vellanikkara	1998	843(E), 21.09.1998
Phule Green Gold (RHRBG-5)	MPKV, Rahuri	2001	348(E), 19.04.2001
Phule Priyanka (RHRBG-1)	MPKV, Rahuri	2001	348(E), 19.04.2001
Pant Karela-2 (PBIG-1)	GBPUAT, Pantnagar	2006	2035(E), 28.11.2006
Pumpkin			
CO-2	TNAU, Coimbtore	1978	13, 19.12.1978
Ambili	KAU, Vellannikkara	1988	1135(E), 01.12.1988
Pusa Vishwas (Selection-107)	IARI, New Delhi	1990	386(E), 15.05.1990
Arka Chandan (IIHR-105-1-1)	IIHR, Bangalore	1990	386(E), 15.05.1990
Sooraj (CM-350)	KAU, Vellanikkara	2003	448(E), 21.04.2003
Kashi Harit (IVPK-226)	IIVR, Varanasi	2006	597(E), 25.4.2006

Crop/ Variety	Deveoping Centre	Year	DAC Notification Number[§] and Date
Narendra Agrim (NDPK-24)	NDUAT, Faizabad	2006	597(E), 25.04.2006
Azad Pumpkin-1 (KPS-1)	CSAUAT, Kanpur	2006	597(E), 25.04.2006
Narendra Amrit	NDUAT, Faizabad	2006	597(E), 25.04.2006
Saras	KAU, Vellanikkara	2006	597(E), 25.04.2006
Narendra Abhushan (NDPKH-1)	NDUAT, Faizabad	2006	597(E), 25.04.2006
Squash			
Early Yellow Prolific	IARI, RS, Katrain	1975	440, 21.08.1975
Punjab Chappankaddoo-1	PAU, Ludhiana	1983	499(E), 08.07.1983
Patty Pan	IIHR, Bangalore	1986	867(E), 26.11.1986
Bottle Gourd			
Pusa Summer Prolific Long (PSPL)	IARI, New Delhi	1975	440, 21.08.1975
Pusa Summer Prolific Round (PSPR)	IARI, New Delhi	1975	440, 21.08.1975
Singapuri Long	Introduction	1975	440, 21.08.1975
Bottle Gourd (Kind)	–	1975	441(E), 21.08.1975
Punjab Round	PAU, Ludhiana	1978	13, 19.12.1978
Kalyanpur Long Green	CSAUAT, Kanpur	1983	2(E), 03.01.1983
CO-1	TNAU, Coimbatore	1984	596(E), 13.08.1984

Crop/ Variety	Deveoping Centre	Year	DAC Notification Number[§] and Date
Arka Bahar	IIHR, Bangalore	1986	867(E), 26.11.1986
Rajendra Chamatkar	RAU, Pusa	1984	636(E), 02.09.1984
Pant Shankar Lauki-1	GBPUAT, Pantnagar	1999	1052(E), 26.10.1999
Punjab Long (LC-2-1)	PAU, Ludhiana	1999	1052(E), 26.10.1999
Narendra Sankar Lauki-4 (NDBGH-4)	NDUAT, Faizabad	2006	597(E), 25.04.2006
Narendra Dharidar (NDBG-208-1)	NDUAT, Faizabad	2006	597(E), 25.04.2006
Narendra Rashmi	NDUAT, Faizabad	2006	597(E), 25.04.2006
Kashi Ganga (DVBG-1)	IIVR, Varanasi	2006	597(E), 25.04.2006
Kashi Bahar (VRH-1)	IIVR, Varanasi	2006	597(E), 25.04.2006
Narendra Jyoti (NDBG-104)	NDUAT, Faizabad	2006	597(E), 25.04.2006
Lauki Azad Shakar-1	CSAUAT, Kanpur	2006	597 (E), 25.04.2006
Pant Lauki-3 (PBOG-61)	GBPUAT, Pantnagar	2006	2035(E), 28.11.2006
Pant Sankar Lauki-2	GBPUAT, Pantnagar	2006	2035(E), 28.11.2006
Muskmelon			
Hara Madhu	PAU, Ludhiana	1970	2035(E), 20.02.1970
Hara Madhu	PAU, Ludhiana	1975	440, 21.08.1975
Hara Madhu	PAU, Ludhiana	1978	1004, 23.03.1978

Crop/ Variety	Deveoping Centre	Year	DAC Notification Number[§] and Date
Durgapura Madhu	ARS, Durgapura	1974	566(E), 21.09.1974
Lucknow Safeda	PAU, Ludhiana	1975	440, 21.08.1975
Muskmelon (Kind)	-	1975	441(E), 21.08.1975
Pusa Sharabati	IARI, New Delhi	1978	1004(E), 23.03.1978
Punjab Sunehri	PAU, Ludhiana	1978	13, 19.12.1978
Gujrat Musk-melon-1	GAU, Anand	1984	596(E), 13.08.1984
Gujrat Musk-melon-2	GAU, Anand	1984	596(E), 13.08.1984
Arka Rajhans	IIHR, Bangalore	1986	867(E), 26.11.1986
Arka Jeet	IIHR, Bangalore	1986	867(E), 26.11.1986
Punjab Hybrid-1	PAU, Ludhiana	1987	834(E), 18.09.1987
Pusa Rasraj	IARI, New Delhi	1993	617(E), 17.08.1993
Punjab Rasila (MR-12)	PAU, Ludhiana	1996	115(E), 10.02.1996
MH-10 (MHI-10)	PAU, Ludhiana	1996	115(E), 10.02.1996
RM-43	RARS, Durgapura	1997	98(E), 08.02.1997
MHY-3	RARS, Durgapura	1999	1052(E), 26.10.1999
RM-50	RARS, Durgapura	2004	642(E), 31.05.2004
MHY-5	RARS, Durgapura	2004	642(E), 31.05.2004
Gujarat Musk-melon-3 (GMM-3)	GAU, Anand	2006	597(E), 25.04.2006
Narendra Musk-melon-1(NDM-2)	NDUAT, Faizabad	2006	597(E), 25.04.2006
Kashi Madhu (IVMM-3)	IIVR, Varanasi	2006	2035(E), 28.11.2006

Crop/ Variety	Deveoping Centre	Year	DAC Notification Number[§] and Date
Watermelon			
Charlston Grey	NBPGR, New Delhi	1975	440, 21.08.1975
Sugarbaby	IARI, New Delhi	1975	440, 21.08.1975
Sugarbaby	IARI, New Delhi	1978	1004, 23.03.1978
Watermelon (Kind)	–	1975	441(E), 21.08.1975
Durgapura Kesar	RARS, Durgapura	1978	13, 19.12.1978
Durgapura Meetha	RARS, Durgapura	1978	13, 19.12.1978
Shipper	PAU, Ludhiana	1978	13, 19.12.1978
Arka Jyoti (Hybrid)	IIHR, Bangalore	1986	867(E), 26.11.1986
Arka Manik	IIHR, Bangalore	1990	386(E), 15.05.1990
Durgapura Lal (RW-177-3)	RARS, Durgapura	2004	642(E), 31.05.2004
Ash Gourd			
CO-2	TNAU, Coimbatore	1984	596(E), 13.08.1984
Pusa Ujjwal (DAG-1)	IARI, New Delhi	2004	642(E), 31.05.2004
Kashi Dhawal (IVAG-502)	IIVR, Varanasi	2006	597(E), 25.04.2006
Kashi Ujjwal (IVAG-90)	IIVR, Varanasi	2006	2035(E), 28.11.2006
Cucumber			
Japanese Long Green	IARI, Katrain	1975	440, 21.08.1975
Solan Green	YSPUHF, Solan	1975	440, 21.08.1975
Kaliyanpur Green	CSAUAT, Kanpur	1983	2(E), 03.01.1983

Crop/ Variety	Deveoping Centre	Year	DAC Notification Number[§] and Date
Sheetal	MPKV, Rahuri	1989	915(E), 06.11.1989
Phule Shubhangi (Sel-75-1-10)	MPKV, Rahuri	2001	348(E), 19.04.2001
Swarna Poorna (CH-20)	HARP, Ranchi	2004	642(E), 31.05.2004
Gujarat Cucumber-1	GAU, Anand	2006	597(E), 25.04.2006
Swarna Ageti	HARP, Ranchi	2006	2035(E), 28.11.2006
Pant Khira-1	GBPUAT, Pantnagar	2006	2035(E), 28.11.2006
Swarna Sheetal	HARP, Ranchi	2006	2035(E), 28.11.2006
Longmelon			
Arka Sheetal	IIHR, Bangalore	1986	867(E), 26.11.1986
Punjab Longmelon No.-1 (H-10)	PAU, Ludhiana	1996	115(E), 10.02.1996
Pointed Gourd			
Rajendra Parwal-1	RAU, Pusa	1994	636(E), 02.09.1994
Rajendra Parwal-2	RAU, Pusa	1994	636(E), 02.09.1994
Swarna Rekha	HARP, Ranchi	2006	2035(E), 28.11.2006
Swarna Alaukik	HARP, Ranchi	2006	2035(E), 28.11.2006
Ivy Gourd			
Sulabha	KAU, Vellanikkara	2006	597 (E), 25.04.2006
Indira Kundru-35	IGKV, Raipur	2006	2035(E), 28.11.2006
Indira Kundru-05	IGKV, Raipur	2006	2035(E), 28.11.2006

Crop/ Variety	Deveoping Centre	Year	DAC Notification Number§ and Date
Ridge Gourd			
CO-1	TNAU, Coimbatore	1978	13, 19.12.1978
Punjab Sadabahar*	PAU, Ludhiana	1978	13, 19.12.1978
Punjab Sadabahar*	PAU, Ludhiana	1991	563(E), 30.08.1991
PKM-1	TNAU, Coimbatore	1984	596(E), 13.08.1984
CO-2	TNAU, Coimbatore	1985	540(E), 24.07.1985
Konkan Harita (DPL-RG-1)	KKV, Dapoli	1996	115(E), 10.02.1996
Arka Sumeet (IIHR-7)	IIHR, Bangalore	2005	664(E), 11.05.2005
Deepthi	KAU, Vellanikkara	2006	597(E), 25.04.2006
Arka Sujat	IIHR, Bangalore	2006	2035(E), 28.11.2006
Swarna Manjari	HARP, Ranchi	2006	2035(E), 28.11.2006
Swarna Uphar	HARP, Ranchi	2006	2035(E), 28.11.2006
Sponge Gourd			
Pusa Chikni	IARI, New Delhi	1975	440, 21.08.1975
Kalyanpur Hari Chikni	CSAUAT, Kanpur	1983	2(E), 03.01.1983
Pusa Supriya (Selection-99)	IARI, New Delhi	1997	98(E), 08.02.1997
Phule Prajakta (GFE-SMG-108)	MPKV, Rahuri	2001	348(E), 19.04.2001
Pusa Sneha	IARI, New Delhi	2004	642(E), 31.05.2004
Gujarat Sponge Gourd-1 (JSGL-55)	GAU, Anand	2006	597(E), 25.04.2006

Crop/ Variety	Deveoping Centre	Year	DAC Notification Number[§] and Date
Swarna Prabha (CHSG-1)	HARP, Ranchi	2006	597(E), 25.04.2006
Azad Toria Chikni-1	CSAUAT, Kanpur	2006	597(E), 25.04.2006
Round Gourd (Tinda)			
Tinda (Kind)	NBPGR, New Delhi	1975	441(E), 21.08.1975
Punjab Tinda (S-48)	PAU, Ludhiana	1978	13, 19.12.1978
Arka Tinda	IIHR, Bangalore	1986	867(E), 26.11.1986
Hisar Tinda	HAU, Hisar	2006	597(E), 25.04.2006
Snake Gourd			
CO-1	TNAU, Coimbatore	1978	13, 19.12.1978
PKM-1	TNAU, Coimbatore	1982	19(E), 14.01.1982
CO-2	TNAU, Coimbatore	1986	867(E), 26.11.1986
Konkan Shweta	KKV, Dapoli	1994	636(E), 02.09.1994
Baby	KAU, Vellanikkara	2006	597(E), 25.04.2006
Manushree	KAU, Vellanikkara	2006	597(E), 25.04.2006
Snapmelon			
Pusa Shandar (DSM-1)	IARI, New Delhi	2004	642(E), 31.05.2004
ROOT VEGETABLES			
Carrot			
No.-29	PAU, Ludhiana	1973	361(E), 30.06.1973
Nantes	IARI, New Delhi	1975	440, 21.08.1975

Crop/ Variety	Deveoping Centre	Year	DAC Notification Number§ and Date
Nantes	IARI, New Delhi	1973	361(E), 30.06.1973
Nantes	IARI, New Delhi	1985	540(E), 24.07.1985
Carrot (Kind)	–	1975	441(E), 21.08.1975
Early Nantes	IARI, New Delhi	1978	13, 19.12.1978
Early Nantes	IARI, New Delhi	1985	540(E), 24.07.1985
Selection-233*	PAU, Ludhiana	1978	13, 19.12.1978
Selection-233*	PAU, Ludhiana	1991	563(E), 30.08.1991
Selection-21	PAU, Ludhiana	1987	165(E), 06.03.1987
Pusa Yamdagni	IARI, Katrain	1988	10(E), 01.01.1988
Pusa Meghali (Sel-5)	IARI, Katrain	1994	636(E), 02.09.1994
Ooty-1	TNAU, Coimbatore	1999	1052(E), 26.10.1999
Shalimar Carrot-1 (SKAU-C-50)	SKUAST, Srinagar	2006	2035(E), 28.11.2006
Radish			
S-White	IARI, RS, Katrain	1973	361(E), 30.06.1973
Japanese White	IARI, RS, Katrain	1973	361(E), 30.06.1973
Japanese White	IARI, RS, Katrain	1975	440, 21.08.1975
White Icicle	IARI, RS, Katrain	1973	361(E), 30.06.1973
Chinese Pink	IARI, RS, Katrain	1975	441(E), 21.08.1975
Nadauni	YSPUHF, Solan	1975	441(E), 21.08.1975
Radish (Kind)	–	1975	441(E), 21.08.1975
Punjab Safaid	PAU, Ludhiana	1978	13, 19.12.1978
Kaliyanpur No.-9	CSAUAT, Kanpur	1983	2(E), 03.01.1983

Crop/ Variety	Deveoping Centre	Year	DAC Notification Number[§] and Date
CO-1	TNAU, Coimbatore	1984	596(E), 13.08.1984
Pusa Chetaki	IARI, New Delhi	1988	10(E), 01.01.1988
Punjab Ageti	PAU, Ludhiana	1988	10(E), 01.01.1988
Arka Nishant	IIHR, Bangalore	1991	527(E), 16.08.1991
Kashi Sweta (IIVR-1)	IIVR, Varanasi	2006	597(E), 25.04.2006
Kashi Hans (VR-2)	IIVR, Varanasi	2006	597(E), 25.04.2006
Hisar Sel-1	HAU, Hisar	2006	597(E), 25.04.2006
Beetroot			
Beetroot (Kind)	–	1975	441(E), 21.08.1975
Detroit Dark Red	IARI, RS, Katrain	1978	13, 19.12.1978
Detroit Dark Red	IARI, RS, Katrain	1985	540(E), 24.07.1985
Crimson Globe	IARI, RS, Katrain	1978	13, 19.12.1978
Crimson Globe	IARI, RS, Katrain	1985	540(E), 24.07.1985
Turnip			
4-White	IARI, RS, Katrain	1973	361(E), 30.06.1973
4-Red	IARI, RS, Katrain	1973	361(E), 30.06.1973
Snowball	IARI, RS, Katrain	1973	361(E), 30.06.1973
Snowball	IARI, RS, Katrain	1975	440, 21.08.1975
Snowball	IARI, RS, Katrain	1978	13, 19.12.1978
Purple Top	IARI, RS, Katrain	1975	440, 21.08.1975
Purple Top	IARI, RS, Katrain	1978	13, 19.12.1978
Golden Ball	IARI, RS, Katrain	1975	440, 21.08.1975

Crop/ Variety	Deveoping Centre	Year	DAC Notification Number[§] and Date
Golden Ball	IARI, RS, Katrain	1978	13, 19.12.1978
Turnip (Kind)	IARI, RS, Katrain	1975	441(E), 21.08.1975
L-1	IARI, RS, Katrain	1978	13, 19.12.1978
LEAFY VEGETABLES			
Amaranthus			
CO-2	TNAU, Coimbatore	1980	2103, 12.08.1980
CO-3	TNAU, Coimbatore	1984	596(E), 13.08.1984
Annapurna	KAU, Vellannikkara	1986	258(E), 14.05.1986
PRA-2 (PRA-9101)	–	2001	92(E), 02.02.2001
Konkan Durangi	KKV, Dapoli	2004	642(E), 31.05.2004
Arka Suguna	IIHR, Bangalore	2006	2035(E), 28.11.2006
Spinach			
Palak (Kind)	NBPGR, New Delhi	1975	441(E), 21.08.1975
Punjab Green	PAU, Ludhiana	1998	843(E), 21.09.1998
Ooty-1	TNAU, Coimbtore	1999	1092(E), 26.10.1999
Arka Anupama	IIHR, Bangalore	2006	597(E), 25.04.2006
OTHER VEGETABLE			
Okra			
Pusa Sawani	IARI, New Delhi	1969	4045, 24.09.1969
Pusa Sawani	IARI, New Delhi	1970	716, 20.02.1970
Pusa Sawani	IARI, New Delhi	1973	361(E), 30.06.1973
Pusa Sawani	IARI, New Delhi	1975	440, 21.08.1975

Crop/ Variety	Deveoping Centre	Year	DAC Notification Number[§] and Date
Pusa Sawani	IARI, New Delhi	1976	786, 02.02.1976
Pusa Sawani	IARI, New Delhi	1985	540(E), 24.07.1985
P-13*	PAU, Ludhiana,	1973	361(E), 30.06.1973
P-13*	PAU, Ludhiana,	1991	563(E), 30.08.1991
S-13	PAU, Ludhiana,	1975	440, 21.08.1975
Bhindi (Kind)	–	1975	441(E), 21.08.1975
CO-2	TNAU, Coimbatore	1978	13, 19.12.1978
Punjab Padmini	PAU, Ludhiana	1983	499(E), 08.07.1983
MDU-1	TNAU, Coimbatore	1985	540(E), 24.07.1985
Gujrat Bhindi-1	GAU, Anand	1986	258(E), 14.05.1986
Selection-2	IARI, New Delhi	1986	258(E), 14.05.1986
Parbhani Kranti	MAU, Parbhani	1986	867(E), 26.11.1986
Punjab-7 (P-7)	PAU, Ludhiana	1988	471(E), 05.05.1988
Utkal Gaurav	OUAT, Bhubaneshwar	1994	636(E), 02.09.1994
Varsa Uphar	HAU, Hisar	1996	115(E), 10.02.1996
Hissar Unnnat	HAU, Hisar	1997	98(E), 08.02.1997
CO-3 (Hybrid-8)	TNAU, Coimbatore	1999	1052(E), 26.10.1999
Phule Kirti (RHROH-4)	MPKV, Rahuri	2001	348(E), 19.04.2001
Shitla Uphar (DVR-1)	IIVR, Varanasi	2001	1135(E), 15.11.2001
Shitla Jyoti (DVR-2)	IIVR, Varanasi	2001	1135(E), 15.11.2001

Crop/ Variety	Deveoping Centre	Year	DAC Notification Number§ and Date
Azad Bhindi-1	CSAUAT, Kanpur	2001	1135(E), 15.11.2001
Gujrat Okra-2	GAU, Anand	2001	1135(E), 15.11.2001
Kashi Pragati (VRO-6)	IIVR, Varanasi	2006	597(E), 25.04.2006
Kashi Vibhuti (VRO-5)	IIVR, Varanasi	2006	597(E), 25.04.2006
Kashi Satdhari (IIVR-10)	IIVR, Varanasi	2006	597(E), 25.04.2006
Phule Utkarsh (GK-IV-3-3-3)	MPKV, Rahuri	2006	597(E), 25.04.2006
HBH-142 (Hybrid)	CCSHAU, Hisar	2006	597(E), 25.04.2006
Kashi Bhairav (DVR-3)	IIVR, Varanasi	2006	597(E), 25.04.2006
Azad Bhindi-2	CSAUAT, Kanpur	2006	597(E), 25.04.2006
Susthira (AE-286-1)	KAU, Vellanikkara	2006	597(E), 25.04.2006
Arka Anamika (IIHR10)	IIHR, Bangalore	2006	2035(E), 28.11.2006
Arka Abhay (IIHR-4)	IIHR, Bangalore	2006	2035(E), 28.11.2006
Hisar Naveen (HRB-107-4)	HAU, Hisar	2006	2035(E), 28.11.2006
Kashi Leela (IIVR-11)	IIVR, Varanasi	2006	2035(E), 28.11.2006

§Varieties notified prior to 15.05.1990 are notified for indefinite period unless denotified, whereas the varieties notified on or after 15.05.1990 are notified for 15 years

*Denotified varieties.

Number of Notified Varieties/Hybrids of Vegetable Crops

Group/Crop	Total	Group/ Crop	Total
Solanaceous		**Cucurbitaceous**	
Tomato	65	Bitter gourd	14
Brinjal	75	Bottle gourd	20
Chilli	45	Cucumber	10
Sweet pepper	10	Pumpkin	11
Leguminous		Squash	03
Vegetable pea	21	Muskmelon	21
Cowpea	14	Watermelon	09
French bean	13	Ridge gourd	10
Indian bean	07	Sponge gourd	08
Bulbs		Ash gourd	04
Onion	29	Pointed gourd	04
Garlic	08	Ivy gourd	03
Cole crops		Snapmelon	01
Cauliflower	22	Longmelon	02
Cabbage	11	Tinda	04
Broccoli	03	Snake gourd	06
Knol khol	07	**Leafy vegetables**	
Root vegetables		Amranthus	06
Carrot	10	Spinach	04
Radish	15	**Other**	
Beetroot	03	Okra	31
Turnip	07		

Varieties/Hybrids Identified through AICRP-VC (1975-2008)

Crop/Variety/ Hybrid	Developing Centre	Year	Re-commended Zone	Remark
	SOLANACEOUS VEGETABLES			
Tomato				
Variety				
S-12	PAU, Ludhiana	1975	-	Small fruited
Pusa Ruby	IARI, New Delhi	1975	-	Small fruited
HS-101	HAU, Hisar	1975	-	Determinate
SL-120	IARI, New Delhi	1975	-	Large fruited
Sweet-72	Gwalior	1975	-	Large fruited
T-1	CSAUAT, Kanpur	1975	-	Large fruited
Pusa Early Dwarf	IARI, New Delhi	1977	-	Small fruited
Sioux	IARI, New Delhi	1977	-	Small fruited
Sel-12	PAU, Ludhiana	1977	-	Small fruited
Punjab Chhuhara	PAU, Ludhiana	1977	-	Small fruited
KS-2	CSAUAT, Kanpur	1985	IV	Determinate
AC-238	GBPUAT, Pantnagar	1985	-	Indeterminate
CO-3	TNAU, Coimbatore	1987	I, IV, V, VI, VII	Determinate
Punjab Kesari	PAU, Ludhiana	1987	I, IV, V, VI	Determinate

Crop/Variety/ Hybrid	Developing Centre	Year	Re-commended Zone	Remark
La-Bonita	NBPGR, New Delhi	1987	I, IV, V, VII	Determinate
Pant T-3	GBPUAT, Pantnagar	1987	I, II, IV, V, VI, VII, VII	Indeterminate
Arka Vikas	IIHR, Bangalore	1987	I, IV, V, VI, VII, VIII	Indeterminate
Arka Saurabh (Sel-4)	IIHR, Bangalore	1987	I, IV, V, VI, VII, VIII	Indeterminate
Sel-7	HAU, Hisar	1990	I, IV, V, VII, VIII	Determinate
Sel-1-6-4	PAU, Ludhiana	1995	I	Determinate
Sel- 32	HAU, Hisar	1996	II, VI, VII	Determinate
DT-10	IARI, New Delhi	1996	IV, VI	Indeterminate
BT-12	OUAT, Bhubaneshwar	1996	I, IV	Indeterminate
KS-17	CSAUAT, Kanpur	1998	IV	Determinate
BT-116-3-2	OUAT, Bhubaneshwar	2001	V, VI	Determinate
NDT-3	NDUAT, Faizabad	2001	IV, VII	Determinate
KS-118	CSAUAT, Kanpur	2001	IV	Determinate
DVRT-2	IIVR, Varanasi	2001	VI	Determinate
BT-20-2-1	OUAT, Bhubaneshwar	2001	IV, VII	Indeterminate
NDT-9	NDUAT, Faizabad	2001	IV	Indeterminate

Crop/Variety/ Hybrid	Developing Centre	Year	Re-commended Zone	Remark
NDTS-2001-3	NDUAT, Faizabad	2004	IV	Determinate
Mani Laima	ICARNEH, Barapani	2004	III	Determinate
IIVR Sel-1	IIVR, Varanasi	2005	V, VII	Determinate
BT-136	OUAT, Bhubaneshwar	2005	II, IV	Determinate
VLT-32	VPKAS, Almora	2005	IV	Determinate
IIVR Sel-2	IIVR, Varanasi	2005	IV	Indeterminate
Hybrid				
ARTH-3	Ankur Seeds, Nagpur	1992	II, VII, VIII	Determinate
ARTH-4	Ankur Seeds, Nagpur	1992	IV, VIII	Indeterminate
MTH-6 (Gulmohar)	Mahyco Seeds, Jalana	1992	VII, VIII	Indeterminate
Pusa Hybrid-2	IARI, New Delhi	1993	I, IV, VI, VII	Determinate
NA-501	Nath Seeds, Aurangabad	1995	IV, VII	Determinate
DTH-4	IARI, New Delhi	1995	VII	Determinate
KT-4	IARI, RS, Katrain	1995	IV	Indeterminate
NA-601	Nath seeds, Aurangabad	1995	VI, VII	Indeterminate
BSS-20	Beejo Sheetal, Jalna	1996	IV, VI, VII	Indeterminate
Avinash-2	Novartis Seeds, Pune	1998	VI	Determinate
HOE-303	Novartis Seeds, Pune	1998	IV	Indeterminate
Sun-496	Sungro Seeds, New Delhi	1999	IV, II, VII	Indeterminate

Crop/Variety/ Hybrid	Developing Centre	Year	Re-comm-ended Zone	Remark
BSS-20	Beejo Sheetal, Jalna	2001	All	Indeterminate
DTH-8	IARI, New Delhi	2001	IV	Determinate
CHTH-1	IARI, New Delhi	2001	IV	Determinate
ARTH-128	Ankur Seeds, Nagpur	2001	VII	Indeterminate
KTH-2	CSAUAT, Kanpur	2002	IV, V	Indeterminate
JKTH-3055	JK Seeds, New Delhi	2004	I, IV	Determinate
KTH-1	CSAUAT, Kanpur	2004	IV	Determinate
Nun-7730	Nunhems Seeds, Bangalore	2004	I, IV	Indeterminate
TH-01462	Syngenta Seeds, Pune	2005	I, II, IV, VI, VII	Determinate
ARTH-734	Ankur Seeds, Nagpur	2008	VIII	Indeterminate
HTH-5	HARP, Ranchi	2008	I	Determinate
Resistant				
BWR-5 (Arka Alok)	IIHR, Bangalore	1992	II	Bacterial wilt
FMH-1 (Arka Vardhan)	IIHR, Bangalore	1993	V, VII	Bacterial wilt
FMH-2	IIHR, Bangalore	1995	IV	Nematode
BT-10	OUAT, Bhubaneshwar	1995	V,VI	Bacterial wilt
H-24	IIVR, Varanasi	1997	V	TLCV
BRH-2	IIHR, Bangalore	1998	VII	Bacterial wilt
LE-415	KAU, Vellanikkara	2004	I, V, VIII	Bacterial wilt
H-86	IIVR, Varanasi	2005	I, IV, V, VIII	TLCV

Crop/Variety/ Hybrid	Developing Centre	Year	Re-comm-ended Zone	Remark
Brinjal				
Variety				
Pusa Purple Long	IARI, New Delhi	1975	IV, VI, VII, VIII	Long
Pusa Purple Cluster	IARI, New Delhi	1975	IV, V, VI, VII	Long
Pusa Kranti	IARI, New Delhi	1977	IV	Long
PB-129-5 (Pant Rituraj)	GBPUAT, Pantnagar	1981	IV	Long
Pant Samrat	GBPUAT, Pantnagar	1981	IV, V, VI	Long
Arka Sheel	IIHR, Bangalore	1981	VIII	Long
Azad Kranti	CSAUAT, Kanpur	1983	IV	Long
Azad Kranti	CSAUAT, Kanpur	1983	I, VI	Long
PB-91-2 (Pant Samrat)	GBPUAT, Pantnagar	1985	–	Round
ARU-1	DARL, Pithoragarh	1985	I	Long
T-3	CSAUAT, Kanpur	1975	–	Round
H-4	HAU, Hisar	1987	–	Long
ARU-2C	DARL, Pithoragarh	1987	I, IV, VI, VIII	Long
K-202-9	GAU, Anand	1987	VI	Round
Aruna	PDKV, Akola	1988	VII	Small round
H-7	HAU, Hisar	1990	IV, VI	Long

Crop/Variety/ Hybrid	Developing Centre	Year	Re-commended Zone	Remark
NDB-25	NDUAT, Faizabad	1990	II, IV, VII	Long
H-8	HAU, Hisar	1990	II, IV, V, VI	Round
BB-26	OUAT, Bhubaneshwar	1993	V	Long
Punjab Barsati	PAU, Ludhiana	1993	IV	Long
Sel-4 (Gulabi)	ANGRAU, Hyderabad	1995	V	Long
DBSR-31	IARI, New Delhi	1995	VI	Long
KS-224	CSAUAT, Kanpur	1995	I, II, IV	Round
DBR-8	IARI, New Delhi	1995	VI	Round
DBSR-44	IARI, New Delhi	1995	VI	Small round
AB-1	GAU, Anand	1996	III, VI, VII	Round
PLR-1	TNAU, Coimbatore	1996	IV, VI, VII	Small round
BB-26	OUAT, Bhubaneshwar	1996	V,VII	Long
BB-13	OUAT, Bhubaneshwar	1997	VIII	Long
KS-331	CSAUAT, Kanpur	1998	IV, V	Long
JB-15	JNKVV, Jabalpur	1998	I	Long
CHBR-1	HARP, Ranchi	1998	IV	Round
DBSR-91	IARI, New Delhi	1998	VII	Small round
JB-64-1-2	JNKVV, Jabalpur	1998	VII	Small round
Green Long	RAU, Samastipur	1998	IV	Green
Punjab Sadabahar	PAU, Ludhiana	2001	IV, VI	Long

Crop/Variety/ Hybrid	Developing Centre	Year	Re-commended Zone	Remark
NDB-28-2	NDUAT, Faizabad	2001	IV	Long
DBL-21 (Pusa Shyamala)	IARI, New Delhi	2004	IV	Long
KS-235	CSAUAT, Kanpur	2004	IV, V, VII	Round
ABSR-2	GAU, Anand	2004	VII	Small long
HABR-4	HARP, Ranchi	2005	IV	
IVBR-1	IIVR, Varanasi	2005	IV	Round
HABL-1	HARP, Ranchi	2006	I	Long
PB-66	GBPUAT, Pantnagar	2007	VII, IV	Long
Hybrid				
Arka Kusumakar	IIHR, Bangalore	1981	VIII	Long
Arka Navneet	IIHR, Bangalore	1981	–	Round
Kat-4	IARI, RS, Katrain	1987	VIII	Long
Pusa Hybrid-6	IARI, New Delhi	1990	IV	Round
Pusa Hybrid-5	IARI, New Delhi	1992	IV, VII, VIII	Long
ARBH-201	Ankur Seeds, Nagpur	1993	IV, V, VI, VII	Long
NDBH-1	NDUAT, Faizabad	1993	IV, VI, VII	Round
ABH-1	GAU, Anand	1993	IV, VI, VII	Small round
MHB-10 (Kalpataru)	Mahyco Seeds, Jalna	1993	IV, VI, VII	Small round

Crop/Variety/ Hybrid	Developing Centre	Year	Re-commended Zone	Remark
MHB-39	Mahyco Seeds, Jalna	1993	IV, VI, VII	Small round
NDBH-6	NDUAT, Faizabad	1995	IV	Long
ABH-2	GAU, Anand	1995	IV, VI	Small round
ABH-2	GAU, Anand	1996	VII	Small round
Phule Hybrid-2	MPKV, Rahuri	1997	VII	Small round
Pusa Hybrid-9	IARI, New Delhi	1997	VI	Round
ARBH-541	Ankur Seeds, Nagpur	2001	All	Long
PBH-6	Pandey Beej, Faizabad	2001	All	Long
JBH-1	GAU, Junagadh	2001	All	Round
BH-1	PAU, Ludhiana	2001	IV	Round
BH-2	PAU, Ludhiana	2002	IV, V	Round
VRBHR-1	IIVR, Varanasi	2002	IV, VI	Round
IVBHL-54	IIVR, Varanasi	2004	IV	Long
ARBH-786	Ankur Seeds, Nagpur	2004	IV	Long
VNR-51	VNR Seed, Raipur	2005	IV, VI	Small round
Navina	VNR Seeds, Raipur	2007	IV	Long
HABH-17	HARP, Ranchi	2007	IV	Round
Resistant				
BB-7 (Utkal Tarini)	OUAT, Bhubaneshwar	1990	II, V	Bacterial Wilt
BWR-12 (Arka Nidhi)	IIHR, Bangalore	1990	VIII	Bacterial Wilt
SM-6-7	OUAT, Bhubaneshwar	1992	VIII	Bacterial Wilt

Crop/Variety/ Hybrid	Developing Centre	Year	Re-commended Zone	Remark
SM-6-6	KAU, Vellanikkara	1996	I, VII, VIII	Bacterial Wilt
BB-44	OUAT, Bhubaneshwar	1996	V, VII	Bacterial Wilt
CHES-309	HARP, Ranchi	2001	I, VII	Bacterial Wilt
BB-64	OUAT, Bhubaneshwar	2004	IV, V, VII, VIII	Bacterial Wilt
Chilli				
Variety				
G-4 (Bhagyamati)	RARS, Lam	1975	–	–
G-5 (Andhra Jyoti)	RARS, Lam	1975	–	–
K-2	TNAU, RS, Kovilpatti	1985	–	–
J-218	JNKVV, Jabalpur	1987	I, IV, V, VI, VII	–
X-235 (LCA-235)	RARS, Lam	1987	I, IV, V, VI, VIII	–
Muslawadi	MPKV, Rahuri	1987	V	–
Sel-1	IIHR, Bangalore	1990	V, VII, VIII	–
LCA-206-B (Prakash)	RARS, Lam	1990	V, VI, VII, VIII	–
AKC-86-39	PDKV, Akola	2001	VII	–

Crop/Variety/ Hybrid	Developing Centre	Year	Re-commended Zone	Remark
BC-14-2	OUAT, Bhubaneshwar	2001	V, VI	–
RHRC-Cluster Erect	MPKV, Rahuri	2001	VII	–
PMR-57/88-K	IIHR, Bangalore	2002	VII	–
LCA-334	RARS, Lam	2002	III, IV, V, VII	–
ASC-2000-02	GAU, Anand	2004	VII	–
KA-2	IIVR, Varanasi	2005	IV	–
LCA-353	RARS, Lam	2007	IV, V, VII	–
BC-225	OUAT, Bhubaneswar	2007	V, VI, VII	–
Hybrid				
HOE-888	Sandoz Seeds, New Delhi	1997	IV, VIII	–
ARCH-236	Ankur Seeds, Nagpur	1997	IV	–
Sungro-86-235	Sungro Seeds, New Delhi	2001	IV, VIII	–
ARCH-228	Ankur Seeds, Nagpur	2002	IV, V, VI	–
CCH-2	IIVR, Varanasi	2005	II, IV, V, VI	–
Sweet Pepper				
Hybrid				
KT-1	IARI, RS, Katrain	1990	I	–
Sel-II	SKUAT, Shalimar	1997	I	–

Crop/Variety/ Hybrid	Developing Centre	Year	Re-comm-ended Zone	Remark
Lario	Syngenta Seeds, Pune	2001	I	–
DARL-202	DARL, Pithoragarh	2002	I, IV	–
KTCPH-3	IARI, RS, Katrain	2005	I, VI, VII	–
LEGUME VEGETABLES				
Garden Pea				
Bonneville	IARI, New Delhi	1975	–	Mid group
GC-141 (JC-141)	JNKVV, Gwalior	1975	–	Mid group
GL-195	JNKVV, Gwalior	1975	–	Mid group
Arkel	IARI, New Delhi	1975	–	Early group
Early December	JNKVV, Gwalior	1977	–	Early group
IP-3 (Pant Uphar)	GBPUAT, Pantnagar	1985	–	Early group
P-88	PAU, Ludhiana	1985	–	Early group
PM-2	GBPUAT, Pantnagar	1987	I	Early group
Lincoln	IARI, Katrain	1987	I	Early group
VL-3	VPKAS, Almora	1987	I, IV, VI	Early group
VL-7	VPKAS, Almora	1992	IV	Early group
Ageta-6	PAU, Ludhiana	1993	I, IV, VI	Early group
VL-6	VPKAS, Almora	1993	IV	Mid group
PH-1	HAU, Hisar	1993	VII	Mid group

Crop/Variety/ Hybrid	Developing Centre	Year	Re-comm-ended Zone	Remark
PH-1	HAU, Hisar	1995	V	Mid group
NDVP-8	NDUAT, Faizabad	1997	IV	Mid group
NDVP-10	NDUAT, Faizabad	1998	IV	Mid group
VL-8	VPKAS, Almora	1998	I	Mid group
VRP-2	IIVR, Varanasi	2001	VI	Early group
NDVP-12	NDUAT, Faizabad	2001	IV	Early group
VRP-3	IIVR, Varanasi	2001	I	Mid group
Oregan Sugar	PAU, Ludhiana	2001	VI	Edible podded
VRP-5	IIVR, Varanasi	2005	I, IV, VIII	Early group
CHP-2	HARP, Ranchi	2005	IV, VI	Mid group
VP-101	VPKAS, Almora	2007	I, IV	Early group
PC-531	PAU, Ludhiana	2007	I, VI, VII	Mid group
Resistant				
PRS-4	CSAUAT, Kanpur	1990	IV, VI, VII	Powdery mildew
JP-4	JNKVV, Jabalpur	1990	IV, VIII	Powdery mildew
JP-83	JNKVV, Jabalpur	1992	VII	Powdery mildew
NDVP-4	NDUAT, Faizabad	1995	IV	Powdery mildew
DPP-68	HPKV, Palampur	2001	IV	Powdery mildew

Crop/Variety/ Hybrid	Developing Centre	Year	Re-comm-ended Zone	Remark
KS-245	CSAUAT, Kanpur	2001	IV	Powdery mildew
NDVP-250	NDUAT, Faizabad	2001	V	Powdery mildew
DPP-9411	HPKV, Palampur	2002	I	Powdery mildew
KTP-8	IARI, RS, Katrain	2005	I, IV, V	Powdery mildew
Cowpea				
Sel-1552 (Pusa Komal)	IARI, New Delhi	1983	–	–
Sel-61-B (Arka Garima)	IIHR, Bangalore	1992	VII, VIII	–
Sel-263	PAU, Ludhiana	1992	IV	–
Sel-2-1	NDUAT, Faizabad	1993	IV	–
IIHR-6	IIHR, Bangalore	1998	IV, VII	–
NDCP-13	NDUAT,Faizabad	2002	II, III, IV, VII	–
IVRCP-1	IIVR, Varanasi	2004	IV	–
CHCP-2	HARP, Ranchi	2005	VIII	–
IIVRCP-4	IIVR, Varanasi	2007	IV, V, VII	–
VR-5	IIVR, Varanasi	2008	IV, V, VII	–
Swarna Harita	HARP, Ranch	2008	II, IV, V, VIII	–

Crop/Variety/ Hybrid	Developing Centre	Year	Re-comm-ended Zone	Remark
French Bean				
VL Boni-1	VPKAS, Almora	1985	NEH Region	–
Arka Komal	IIHR, Bangalore	1987	I, VII, VIII	–
UPF-191	GBPUAT, Pantnagar	1987	IV, VIII	–
IIHR-909	IIHR, Bangalore	1997	I	–
CH-812	HARP, Ranchi	2001	VII, III	–
CH-819	HARP, Ranchi	2001	I	–
IVFB-1	IIVR, Varanasi	2005	I, VII	–
Indian Bean				
Deepaliwal	PDKV, Akola	1990	V, VII	–
CHDB-1	HARP, Ranchi	2004	IV	–
		BULB VEGETABLES		
Onion				
Punjab Selection	PAU, Ludhiana	1975	-	More self-life
Pusa Red	IARI, New Delhi	1975	-	More self-life
Pusa Ratnar	IARI, New Delhi	1975	-	Red
S-131	IARI, New Delhi	1977	-	White
N-257-9-1	MPKV, Rahuri	1985	-	White
N-2-4-1	MPKV, Rahuri	1985	-	Red
Pusa Madhavi (Line-120)	IARI, New Delhi	1987	I, IV, VI, VII	*Rabi* season

Crop/Variety/ Hybrid	Developing Centre	Year	Re-commended Zone	Remark
Arka Kalyan	IIHR, Bangalore	1987	IV, VI, VII, VIII	*Rabi* season
Arka Niketan	IIHR, Bangalore	1987	VII	*Kharif* season
Agri Found Dark Red	NHRDF, Nasik	1987	IV	*Kharif* season
VL-3	VPKAS, Almora	1990	IV	*Rabi* season
Punjab Red Round	PAU, Ludhiana	1993	IV	*Rabi* season
Agrifound Light Red	NHRDF, Nasik	1995	VI, VIII	*Rabi* season
PBR-5	PAU, Ludhiana	1997	VI	*Rabi* season
HOS-1	HAU, Hisar	2006	VI	*Rabi* season
B-780-5-2-2	NRCOG, Pune	2007	VI	*Rabi* season
Garlic				
G-41	NHRDF, Karnal	1988	IV, VII	–
G-1	NHRDF, Karnal	1990	IV, VI, VII	–
G-50	NHRDF, Karnal	1993	IV	–
G-282	NHRDF, Karnal	1998	IV, VI, VII	–
VLG-7	VPKAS, Almora	2001	I	–
DARL-52	DARL, Pithoragarh	2002	I	–
G-323	NHRDF, Nasik	2002	VI	–

Crop/Variety/ Hybrid	Developing Centre	Year	Re-commended Zone	Remark
		COLE VEGETABLES		
Cauliflower				
Early Kunwari	PAU, Ludhiana	1975	–	Early group
327-14-8-3	IARI, New Delhi	1975	–	September maturity
351-4-1	IARI, New Delhi	1975	–	October maturity
Improved Japanese	IARI, New Delhi	1975	–	November maturity
EC-12012	IARI, Katrain	1975	–	January maturity
Pusa Snowball	IARI, Katrain	1975	–	January maturity
K-1	IARI, Katrain	1979	–	January maturity
114-S-1	GBPUAT, Pantnagar	1981	–	January maturity
Line 6-1-2-1	IARI, New Delhi	1985	–	December maturity
235-S	GBPUAT, Pantnagar	1990	II, VIII	November maturity
KT-25	IARI, Katrain	2001	I	Snowball group
IVREC-2	IIVR, Varanasi	2005	IV	Early group
IIVRMC-12	IIVR, Varanasi	2008	IV	Mid group
DC-76	IARI, New Delhi	2008	I, VI	Mid group

Crop/variety/ hybrid	Developing centre	Year	Re-comm-ended zone	Remarks
Synthetic				
Synthetic-1	IARI, New Delhi	1975	–	December maturity
Early Synthetic	IARI, New Delhi	1990	IV, VIII	Early group
Hybrid				
Pusa Hybrid-1	IIVR, Varanasi	1992	II, IV	Early group
DCH-541	IARI, New Delhi	2002	II, IV	Early group
SYCFH-202	Syngenta Seeds, Pune	2004	IV, VII	Early group
Summer King	Sungro Seeds, New Delhi	2004	I, IV	Early group
SYCFH-203	Syngenta Seeds, Pune	2005	IV, V, VII	Early group
Cabbage				
Sel-8	IARI, New Delhi	1985	II, IV	Variety
Shri Ganesh Gol	Mahyco Seeds, Jalana	1992	V	Hybrid
Nath-401	Nath Seeds, Aurangabad	1993	I, IV, V, VI, VII	Hybrid
BSS-32	Beejo Sheetal Seeds, Jalana	1995	VII	Hybrid
Nath-501	Nath Seeds, Aurangabad	1997	VII	Hybrid
Quisto	Novartis Seeds, Pune	1998	IV	Hybrid

Crop/variety/ hybrid	Developing centre	Year	Re-comm-ended zone	Remarks
KGMR-1	IARI, RS, Katrain	2005	I, IV	Hybrid
Green Emperor	Tokita Seeds, Bangalore	2007	I	Hybrid
Pusa Synthetic	IARI, RS, Katrain	1992	IV, I, II	Synthetic
ROOT VEGETABLES				
Carrot				
Sel-5 (Pusa Meghali)	IARI, New Delhi	1975	VII	–
Pusa Meghali	IARI, New Delhi	1992	VII	–
SKAUC-50	SKUA&T, Srinagar	2006	I	–
Hybrid-I	Mahyco Seeds, Jalna	1992	I, VII	–
CUCURBITACEOUS VEGETABLES				
Bitter Gourd				
Variety				
Priya	KAU, Vellanikkara	1992	II, VII, VIII	–
RHRBG-4-1	MPKV, Rahuri	1998	IV, VII	–
PBIG-1	GBPUA&T, Pantnagar	2001	IV	–
Hybrid				
Pusa Hybrid-2	IARI, New Delhi	2002	IV, V, VI	–
NBGH-167	Nirmal Seeds, Jalgaon	2004	IV	–
Vivek	Sungro Seed, New Delhi	2008	VIII	–

Crop/variety/ hybrid	Developing centre	Year	Re-commended zone	Remarks
Bottle Gourd				
Variety				
Pusa Naveen	IARI, New Delhi	1992	VII	–
KBG-16	CSAUA&T, Kanpur	1998	IV	–
PBOG-61	GBPUA&T, Pantnagar	2001	IV, VI	–
NDBG-104	NDUA&T, Faizabad	2002	IV	–
NDBG-132	NDUA&T, Faizabad	2004	VI	–
Sel P-6	IARI, New Delhi	2008	IV, VII	–
Hybrid				
PBOG-2	GBPUA&T, Pantnagar	2001	VII	–
PBOG-1	GBPUA&T, Pantnagar	2001	IV	–
NDBH-4	NDUA&T, Faizabad	2001	All	–
Pumpkin				
CM-14	KAU, Vellanikkara	1987	IV, V, VII	–
Pusa Vishwas	IARI, New Delhi	1987	I, VIII, IV, V	–
Arka Chandan	IIHR, Bangalore	1987	VIII	–
Arka Suryamukhi	IIHR, Bangalore	1987	VIII	–
CM-350	KAU, Vellanikkara	2001	VII, VIII	–
NDPK-24	NDUAT, Faizabad	2001	IV, VI	–

Crop/variety/ hybrid	Developing centre	Year	Re-commended zone	Remarks
Cucumber				
Variety				
CHC-2	HARP, Ranchi	2001	IV	–
CH-20	HARP, Ranchi	2001	IV	–
PCUC-28	GBPUAT, Pantnagar	2001	I, VII, VIII	–
Hybrid				
PCUCH-1	GBPUAT, Pantnagar	2001	All	–
Hybrid No.-1	Century Seeds, New Delhi	2004	I, IV, VII	–
PCUCH-3	GBPUAT, Pantnagar	2005	I, IV	–
Muskmelon				
Variety				
Pusa Sharbati	IARI, New Delhi	1975	IV	–
Hara Madhu	PAU, Ludhiana	1975	IV, VII	–
Sl-45 (Pusa Madhuras)	IARI, New Delhi	1975	IV, VI, VIII	–
Arka Rajhans	IIHR, Bangalore	1975	VIII	–
Arka Jeet	IIHR, Bangalore	1975	VIII	–
Durgpura Madhu	ARS, Durgapura	1975	IV	–
MHY-5	ARS, Durgapura	2001	VII	–
NDM-15	NDUAT, Faizabad	2002	IV	–

Crop/variety/ hybrid	Developing centre	Year	Re- comm- ended zone	Remarks
IVMM-3	IIVR, Varanasi	2006	IV	–
GMM-3	GAU, Anand	2008	IV, VII	–
Hybrid				
Hybrid M-3	IARI, New Delhi	1990	IV	–
Resistant				
DMDR-1	IARI, New Delhi	2001	Recommended as a source of resistance for CGMV	
DMDR-2	IARI, New Delhi	2001	Recommended as a source of resistance for DM + CGMV	
Watermelon				
Variety				
Durgapura Meetha	ARS, Durgapura	1975	IV, VI, VII, VIII	–
Sugar Baby	IARI, New Delhi	1975	V, VII, VIII	–
Arka Manik	IIHR, Bangalore	1987	IV, VII, VIII	–
Hybrid				
MHW-6	MAHYCO, Jalna	1999	–	–
Arka Jyoti	IIHR, Bangalore	1981	–	–
Ridge Gourd				
CHRG-1	HARP, Ranchi	2001	IV	–

Crop/variety/ hybrid	Developing centre	Year	Re-comm-ended zone	Remarks
PRG-7	GBPUT, Pantnagar	2001	VII	–
IIHR-7	IIHR, Bangalore	2001	VIII	–
Sponge Gourd				
Sel-99	IARI, New Delhi	1995	IV, VI	–
CHSG-1	HARP, Ranchi	2005	IV	–
JSGL	GAU, Junagarh	2005	VII	–
KSG-14	CSAUAT, Kanpur	2006	IV	–
PSG-40	GBPUAT, Pantnagar	2007	I, VII	–
Ash Gourd				
IVAG-90	IIVR, Varanasi	2006	IV, VIII	–
PAG-72	GBPUAT, Pantnagar	2006	VIII	–
Pusa Ujjwal	IARI, New Delhi	2007	I	–
	OTHER VEGETABLE			
Okra				
Hybrid				
DVR-1	IIVR, Varanasi	1998	IV, VII	–
DVR-2	IIVR, Varanasi	1998	VI	–
DVR-3	IIVR, Varanasi	2001	All	–
DVR-4	IIVR, Varanasi	2001	IV, V, VII	–
HBH-142	HAU, Hisar	2005	IV, V, VII, VII	–
SOH-152	Syngenta, Pune	2005	IV, VII, VII	–

Crop/Variety/ Hybrid	Developing Centre	Year	Re-comm-ended Zone	Remark
SOH-1016	Syngenta, Pune	2007	IV, VII	–
NBH-180	Nuzi Vedu Seeds, Secunderabad	2007	VII	–
JNDOH-02-2	JAU, Junagarh	2008	II, V, VI, VII, VIII	–
Resistant				
P-7	PAU, Ludhiana	1990	All	–
PB-57	MAU, Parbhani	1990	All	–
Sel-10 (Arka Anamika)	IIHR, Bangalore	1990	All	–
Sel-4 (Arka Abhay)	IIHR, Bangalore	1992	II	–
HRB-55 (Varsa Uphar)	HAU, Hisar	1995	VI	–
HRB-9-2	HAU, Hisar	1996	IV, VI	–
VRO-3	IIVR, Varanasi	2001	–	–
VRO-4	IIVR, Varanasi	2001	IV, V	–
VRO-5	IIVR, Varanasi	2002	VI	–
VRO-6	IIVR, Varanasi	2002	IV, V	–
NDO-10	NDUAT, Faizabad	2005	IV	–
HRB-107-4	HAU, Hisar	2005	VI, VIII	–
IIVR-11	IIVR, Varanasi	2005	VI, VII	–
JNDOL-03-1	AAU, Junagarh	2007	VII, VIII	–

Number of Varieties/Hybrids Identified through AICRP-VC (1975-2008)

Crop	Open Pollinated	Hybrid/ Synthetic	Resistant	Total
Tomato	36	24	8	68
Brinjal	46	25	7	78
Chilli	18	10	–	28
Garden pea	25	–	9	34
Cowpea	11	–	–	11
French bean	7	–	–	7
Indian bean	2	–	–	2
Okra	0	9	14	23
Onion	17	–	–	17
Garlic	7	–	–	7
Cauliflower	14	7	–	21
Cabbage	1	8	–	9
Carrot	3	1	–	4
Muskmelon	9	2	3	14
Watermelon	4	1	–	5
Bitter gourd	4	3	–	7
Pumpkin	6	–	–	6
Cucumber	3	3	–	6
Ridge gourd	3	–	–	3
Bottle gourd	5	3	–	8
Sponge gourd	5	–	–	5
Ash gourd	3	–	–	3
	229	**96**	**41**	**366**

Crop-wise Breeder Seed Production (kg) During the Last 5 Years

Variety/ Hybrid	2002 -03	2003 -04	2004 -05	2005 -06	2006 -07
Amaranthus					
CO-2	2.50	2.50	1.00	5.00	1.00
Ash gourd					
CO-2	–	2.00	0.50	–	3.00
CO-1	0.500	–	–	–	0.500
Beet Root					
DDR	–	6.7	–	–	–
Bitter gourd					
Arka Harit	13.00	30.00	12.00	5.50	4.50
Hirkani	7.00	8.00	1.50	1.00	–
Baramasi	20.00	10.00	13.00	3.00	–
Kalyanpur Sona	25.00	25.00	14.00	–	–
Konkan Tara	5.00	5.00	1.50	6.00	–
Punjab-14	6.00	–	–	–	–
Pusa Vishesh	11.00	10.00	5.00	6.00	1.02
Phule Green Gold	–	5.00	3.00	4.00	2.00
BG-14	–	10.00	4.00	10.00	10.00
PDM	–	–	–	10.00	–
Bottle gourd					
Arka Bahar	50.00	50.00	50.00	50.00	1.00

Variety/ Hybrid	2002 -03	2003 -04	2004 -05	2005 -06	2006 -07
Kalyanpur Long Green	10.00	–	–	–	–
Punjab Komal	6.00	–	4.00	27.00	35.00
Punjab Long	0.50	–	–	–	15.00
PSPL	122.00	–	39.00	29.00	32.75
Pusa Naveen	62.00	–	44.00	25.50	8.00
Samrat	4.00	–	4.00	2.00	3.00
Azad Harit	–	20.00	–	–	–
Kalyanpur Long Green	–	15.00	–	–	–
Narendra Dharidar	–	15.00	–	–	–
Pb. Komal	–	6.00	–	–	–
PSPL	–	48.00	–	–	–
Pusa Naveen	–	24.40	–	–	–
Sabour Chamatkar	–	10.00	–	–	–
Samrat	–	5.00	–	–	–
Pusa Hybrid-3 (F)	–	–	–	1.80	3.00
Pusa Hybrid-3 (M)	–	–	–	0.450	1.50
Rajendra Chamatkar	–	–	–	6.00	
Brinjal					
Arka Kusumakar	8.00	5.00	2.00	5.00	0.400
Arka Sheel	4.00	–	–	–	–
Arka Shirish	4.00	4.00	–	–	0.200
Azad Hybrid (F)	0.100	–	–	–	–
Azad Hybrid (M)	0.100	–	–	–	–
Hisar Shyamal	5.20	5.00	10.00	1.00	–

Variety/ Hybrid	2002 -03	2003 -04	2004 -05	2005 -06	2006 -07
Manjri Gota	4.00	2.00	1.00	15.00	0.800
Pant Rituraj	10.00	10.00	10.00	32.00	5.00
Pant Samrat	5.00	50.00	50.00	48.00	50.00
PPC	4.975	9.87	8.625	9.375	10.975
PPL	4.785	4.84	3.385	2.285	1.550
Pusa Bindu	0.250	–	–	–	–
Pusa Hybrid-5 (F)	0.750	1.30	–	–	0.510
Pusa Hybrid-5 (M)	0.200	0.90	–	–	0.135
Pusa Hybrid-6 (F)	0.950	0.800	–	–	0.500
Pusa Hybrid-6 (M)	0.300	0.250	–	–	0.125
Pusa Kranti	1.795	1.40	1.00	1.00	0.510
Pusa Upkar	1.250	2.00	1.40	1.08	0.365
Pusa Ankur	–	–	0.30	–	0.275
Pusa Uttam	–	–	–	–	0.595
Utkal Anushree	–	–	–	–	4.00
Cabbage					
Golden Acre	5.635	4.16	–	0.900	26.75
Pride of India	–	0.100	0.300	0.300	0.250
Capsicum					
California Wonder	5.69	2.445	0.715	0.412	1.090
Carrot					
Pusa Kesar	75.00	45.00	25.00	42.00	25.00
Nantes	–	–	4.50	2.50	1.00
Pusa Meghali	–	–	–	2.00	–

Variety/ Hybrid	2002 -03	2003 -04	2004 -05	2005 -06	2006 -07
Cauliflower					
Pant Subhra	3.00	30.00	–	–	–
PSBK-1	10.850	5.37	–	1.055	15.50
Pusa Deepali	3.90	2.00	–	–	–
Narendra Gobhi-1	–	3.00	–	–	–
Summer squash					
Punjab Chappan Kaddoo-1	5.00	–	–	–	–
Chilli					
CA-960	3.45	3.45	0.400	–	–
CH-1 (F)	0.80	0.800	0.800	1.00	4.00
CH-1 (M)	0.30	0.350	0.350	1.00	1.00
CH-3 (F: MS-12)	0.30	0.300	0.300	1.00	–
CH-3 (M: S-2530)	0.10	0.150	0.150	5.00	–
G-4	4.50	8.25	3.450	3.10	2.50
G-5	0.750	–	–	–	–
Konkan Kirti	14.00	15.00	3.00	7.50	
Pusa Jwala	10.275	–	10.70	20.50	10.00
Pusa Sadabahar	4.50	3.815	14.00	1.90	2.35
Pant C-1	3.00	30.00	50.00	–	1.10
Phule Jyoti	5.00	–	–	0.500	–
Phule-5	2.00			–	
X-235	2.90	2.95	2.95	1.60	1.00
Jawahar Mirch-283	–	15.00	7.50	–	–
Parbhani Tejas	–	0.350	–	–	–
CO-3	–	–	–	–	0.100

Variety/ Hybrid	2002 -03	2003 -04	2004 -05	2005 -06	2006 -07
J-28	–	–	–	–	0.400
Jayanti	–	–	–	–	5.00
RCH-1	–	–	–	–	1.00
Utkal Reshmi	–	–	–	–	0.300
Cowpea					
Pusa Komal	220.00	505.00	163.00	10.00	66.00
Narendra Lobia	–	100.00	50.00	114.00	–
RCV-7	–	6.00	–	110.00	44.50
S-263	–	35.00	10.00	50.00	40.00
Arka Garima	–	–	60.00	50.00	10.00
Gujarat-4	–	–	52.00	–	–
NDCP-13	–	–	–	–	10.00
Utkal Manika	–	–	–	–	70.00
Cucumber					
JLG	8.49	–	–	–	–
Kalyanpur Green	20.00	20.00	30.00	–	–
Poinsette	8.00	–	3.50	–	–
Poona Khira	14.00	10.00	4.50	3.00	3.00
Himangi	–	10.00	12.00	8.00	2.00
Japanese Green	–	10.20	9.50	5.40	7.90
Poinsette	–	10.00	–	1.00	2.40
Shubhangi	–	5.00	1.00	5.00	–
Dolichos	–	–	–	–	–
Konkan Bhushan	60.00	130.00	105.00	110.00	85.00

Crop-wise Breeder Seed Production During the Last 5 Years

Variety/ Hybrid	2002 -03	2003 -04	2004 -05	2005 -06	2006 -07
Arka Vijay	–	15.00	–	–	–
PEP	–	38.00	19.00	–	20.00
Phule Gauri	–	–	–	–	50.00
Rajni	–	–	–	–	40.350
French Bean					
Arka Komal	860.00	1250.00	1150.00	1190.00	840.00
Contender	920.15	1400.15	844.50	444.50	321.00
Pant Anupam	–	300.00	–	–	–
Phule Surekha	–	–	–	–	20.00
Phule Suyash	–	–	–	–	30.00
Arka Suvidha	–	–	–	–	25.00
Longmelon					
Punjab Longmelon	–	–	–	–	11.00
Muskmelon					
Hara Madhu	20.00	39.25	9.25	–	10.00
Madhuras	20.40	15.00	5.00	5.40	–
MS-1	5.00	–	–	–	–
Punjab Sunehri	12.50	15.00	4.25	25.00	16.00
Punjab Hybrid-1 (F)	–	20.00	0.50	20.00	20.00
Punjab Hybrid-1 (M)	–	0.750	0.25	21.00	10.00
Arka Rajhans	–	–	2.00	–	–
Knol Kohl					
White Vienna		1.635	–		0.50

Variety/ Hybrid	2002 -03	2003 -04	2004 -05	2005 -06	2006 -07
Okra					
A-4	114.00	–	–	–	–
Arka Anamika	372.00	160.00	118.00	100.00	129.00
Parbhani Kranti	1655.00	485.00	300.00	213.00	30.00
Punjab Padmini	10.00	20.00	10.00	–	–
Punjab -7	30.00	27.00	1.00	–	24.00
Punjab -8	55.00	50.00	40.00	100.00	60.00
Pusa Sawani	168.00	154.00	–	30.00	10.00
Utkal Gaurav	20.00	25.00	25.00	50.00	30.00
Varsha Uphar	250.00	300.00	348.00	94.00	200.00
Azad Bhindi-1	–	60.00	40.00	70.00	61.00
Hisar Unnat	–	2.00	2.00	–	–
Arka Abhay	–	–	18.00		20.00
Gujarat-2	–	–	10.00	0.500	11.00
VRO-6	–	–	–	45.00	125.00
Phule Utkarsha	–	–	–	–	15.00
Onion					
AFDR	45.00	98.00	90.00	70.500	42.50
AFLR	155.00	89.10	230.00	85.500	75.00
N-53	60.00	15.00	30.00	5.00	12.00
Pusa Red	43.00	32.00	18.00	15.00	–
Kalyanpur Red Round	10.00	15.00	–	–	–
Baswant-780	-	20.00	–	90.00	–
Arka Kalyan	10.00	–	40.00	20.00	–
Arka Niketan	1.00	–	25.00	15.00	–

Variety/ Hybrid	2002 -03	2003 -04	2004 -05	2005 -06	2006 -07
Punjab White	10.00	–	–	–	–
Punjab Red Round	–	–	–	6.00	–
Spinach (Palak)					
All green	300.00	250.00	120.00	100.00	200.00
Kalyanpur Green	20.00	30.00	–	–	–
Jobner Green	–	20.00	7.00	120.00	5.00
Pusa Bharati	–	–	15.00	20.00	30.00
Pusa Harit	–	–	–	12.00	15.00
Pea					
AP-1	5130.00	9200.00	4470.00	2240.00	2700.00
Arkel	25650.00	15620.00	15780.00	6152.00	4950.00
AP-3	8100.00	6350.00	6830.00	6000.00	2950.00
Bonneville	2600.00	1500.00	450.00	250.00	50.00
E-6	840.00	2500.00	2200.00	2000.00	2500.00
Lincoln	275.250	297.75	–	201.75	210.00
Pant Sabji Matar-3	3700.00	3700.00	3100.00	4657.00	4000.00
Punjab-87	–	200.00	–	800.00	–
Punjab-88	–	500.00	430.00	–	430.00
VL-7	–	50.00	100.00	–	–
Azad P-4	–	–	–	–	–
Pusa Pragati	–	200.00	–	392.00	200.00
Pumpkin					
Arka Chandan	10.70	20.00	25.00	40.00	–
Pusa Vishwas	–	7.80	7.30	8.30	2.265

Variety/ Hybrid	2002 -03	2003 -04	2004 -05	2005 -06	2006 -07
Radish					
Arka Nishant	50.00	30.00	10.00	30.00	–
Chinese Pink	4.00	5.00	5.50	6.50	9.00
Japanese White	66.675	101.90	–	45.025	20.10
Kashi Shweta	–	2.50			
Pusa Chetki	85.00	20.00	25.00	14.00	6.50
Pusa Himani	36.850	41.35	–	11.850	11.05
Pusa Reshmi	55.00	15.00	10.00	10.00	5.00
Ridge gourd					
Arka Sujat	5.00	10.00	12.00	–	2.00
Arka Sumit	5.00	–	–	60.00	–
PKM-1	3.00	2.50	16.50	12.00	0.500
Pusa Nasdar	25.00	27.00	13.00	15.50	17.80
Sponge gourd					
Kalyanpur Harichikni	5.00	16.00	5.00	–	0.500
Pusa Chikni	54.00	35.00	17.00	10.00	5.00
Pusa Supriya	60.00	52.00	35.30	10.00	15.00
Tinda					
Arka Tinda	9.50	8.00	–	15.00	24.00
S-48 (Punjab Tinda)	50.00	40.00	19.00	80.00	37.00
Snake gourd					
PKM-1	–	4.00	–	–	–
CO-1	–	–	–	–	0.50

Variety/ Hybrid	2002 -03	2003 -04	2004 -05	2005 -06	2006 -07
Tomato					
Arka Vikas	8.00	12.00	5.00	25.00	1.700
Azad T-3	0.40	–	–	1.50	–
Hisar Lalit	4.80	3.00	6.00	–	–
H-86	0.200	0.250	–	–	–
Punjab Chhuhara	0.50	3.00	–	–	–
PED	9.00	12.390	5.00	7.80	1.50
PKM-1	5.00	6.00	4.00	30.00	2.550
Pusa Divya (F)	0.090	0.105	0.050	–	–
Pusa Divya (M)	0.095	0.100	0.075	–	–
Pusa Gaurav	1.60	–	0.100	2.00	–
Pusa Hybrid-1 (F)	0.400	0.350	0.100	–	–
Pusa Hybrid-1 (M)	0.150	0.100	0.020	–	–
Pusa Hybrid-2 (F)	0.400	0.450	0.300	–	0.350
Pusa Hybrid-2 (M)	0.200	0.150	0.100	–	0.150
Pusa Ruby	18.00	15.120	7.00	3.50	2.00
Pusa Hybrid-4 (F)	–	0.400	0.100	–	0.300
Pusa Hybrid-4 (M)	–	0.150	0.050	–	0.200
Pusa Gaurav	–	1.680	–	–	–
Narendra-2	–	2.50	1.00	1.00	–
Arka Abhijit (F)	–	0.200	–	–	–
Arka Abhijit (M)	–	0.100	–	–	–
Pusa Uphar	–	–	0.100	4.00	–
Pant T-3	–	–	–	–	10.00
Pusa Hybrid-8 (M)	–	–	–	–	0.125

Variety/ Hybrid	2002 -03	2003 -04	2004 -05	2005 -06	2006 -07
Pusa Hybrid-8 (F)	–	–	–	–	0.500
Utkal Deepti	–	–	–	–	5.00
Turnip					
PTWG	8.80	8.65	–	2.20	3.75
Watermelon					
Arka Jyoti (F)	1.00	1.00	–	0.800	–
Arka Jyoti (M)	0.50	0.50	–	0.200	–
Arka Manik	5.00	19.00	10.00	8.70	9.00
Sugar Baby	127.00	89.50	73.00	65.00	43.07
Ayashi Amato	–	16.25	17.25	17.25	–

Centers Involved in Research on Vegetable Crops

Centre	Address of Officer-in-charge	Phone/Fax/Email
IIVR, Varanasi	Director Indian Institute of Vegetable Research, P.B. No. 01, P.O. Jakhini (Shahanshahpur), Varanasi-221 305 (U.P.)	0542-2635236 0542-2635247 05443-229007 pccell@gmail.com
TNAU, Coimbatore	Head Department of Vegetable Crops Hort. College & Res. Institute, TNAU, Coimbatore-641 003 (Tamil Nadu)	0422-2431222 ext. 283 0422-2431672, 2454186, 2445414, 5511283 vegetables@tnau.ac.in
IIHR, Bangalore	Head Division of Vegetable Crops, IIHR, Hessaraghatta Lake Post, Bangalore-560 089 (Karnatka)	080-28466420-23 ext.-224 080-28466291 vegetables@iihr.ernet.in
IARI, RS, Katrain	Head IARI Research Station, Katrain-175 129, Kullu Valley (H.P.)	01902-240124, 241280 01902-240124 headrask@yahoo.co.in
PAU, Ludhiana	Head Department of Vegetable Crops Punjab Agriculture University Ludhiana-141 004 (Punjab)	0161-2401960 ext.-370 2404460 0161-2400945 Abi_1975@rediffmail.com
IARI, New Delhi	Head Division of Vegetable Science IARI, New Delhi-110 012	011-25846628, 25847148 011-25766420, 25847148
MPKV, Rahuri	Sr. Vegetable Breeder Deptt. of Horticulture MPKV, Rahuri-413 722 (M.S.)	02426-243247 02426-243302 hodhort@indiatimes.com
RAU, Sabour	Vegetable Breeder AICRP (Veg. Crops) Bihar Agriculture College Bhagalpur, Sabour-813 210 (Bihar)	0641-245158 0641-2451400

Centre	Address of Officer-in-charge	Phone/Fax/Email
MAU, Parbhani	Research Officer Hort. Research Sation Sub, Campus MAU, Parbhani -431 402 (MS)	02452-223801-08 ext. 4250 aicvipmaupbn@yahoo.com yahoo.com
OUAT, Bhubanes-war	Officer-in-Charge (AICRP-VC) Directorate of Reseach, 2nd Floor, Administrative Building, OUAT, Bhubaneswar-751 003 (Orissa)	0674-2401692 0674-2407780 vegres.ouat@yahoo.com
UAS, Dharwad	Project Leader (AICRP-VC) Department of Horticulture, University of Agricultural Sciences Dharwad-580 005 (Karnataka)	0836-347783/ 347958 0836-348349
RAU, RRS, Durgapura	Officer Incharge (AICRP-VC) Department of Horticulture, Agril Res. Station, ARS Durgapura, Jaipur-302 018 (Rajsthan)	0141-2550391, 2550392 0141-2550229 bsyhortjpr@yahoo.co.in
NDUA&T, Faizabad	Head Department of Vegetable Crops NDUA&T, Faizabad-224 229 (U.P.)	05270-262118, 262023 262076
HAU, Hisar	Head Department of Vegetable Crops, CCSHAU, Hisar-125 004 (Haryana)	01662-233640, 231171 to 231173 01662-234952 vegscience@hau.ernet.in
ANGRAU, Hyderabad	Principal Scientist & Head Department of Vegetable Crop, ANGRAU, Rajendra Nagar, Heyderabad-500 030 (A.P.)	040-24018016 040-24018016 vegreshyd@yahoo.com
JNKVV, Jabalpur	Head Deptt. of Vegetable Crops, College of Agriculture, JNKV, Jabalpur-482 004 (M.P.)	0761-2481771, 2481773 Ext. 345 0761-343382, 42719 ars-jpr@raj.nic.in
AAU, Jorhat	Officer Incharge (AICRP-VC), Department of Horticulture Assam Agriculture University, Jorhat-785 013 (Assam)	0376-320951 0376-320919 jdevi@aau.ac.in

Centre	Address of Officer-in-charge	Phone/Fax/Email
GAU, Junagadh	Officer Incharge (AICRP-VC) Vegetable Research Station, Gujarat Agriculture University, Junagadh-362 001 (Gujarat)	0285-2672080 ext. 372 2621355 k.v.talatiya@jau.in
CSAUAT, Kanpur	Head Deptt. of Vegetable Science CSAUAT, Kalyanpur, Kanpur-208 024 (U.P.)	0512-2580289 0512-2210408
BCKV, Kalyani	Officer Incharge (AICRP-VC) Department of Horticulture, BCKV, Nadia, Kalyani-741 235 (WB)	033-25828407 03473-233275 ivs@ca13.vsnl.net.in
ANGRAU, RARS, Lam	Head, Chilli Section ANGRAU, Reg. Agri. Res. Station, Lam, Guntur-522 034 (AP)	0863-2524017, 2524022 umajyotik@hotmail.com
GBPUA&T, Pantnagar	Head Deptt. of Vegetable Science GBPUA&T, Udham Singh Nagar, Pantnagar-263 145 (U.P.)	05944-235199 05944-233473, 33608
IGKV, Raipur	Head Deptt. of Horticulture, IGKV, Raipur-492 012 (M.P.)	0771-2425219, 2497012 Ext. 136 0771-2442131 glsigau@rediffmail.com
YSPUHF, Solan	Head Deptt. of Vegetable Crops, College of Horticulture, Dr. Y.S. Parmar University of Hort. & Forestry, Solan-173 230 (H.P.)	01792-252329 01792-262242 vgc@yspuhf.hp.nic.in
SKUAT, Srinagar	Head Division of Olericulture, Shere-E-Kashmir Univ. of Agril. Sci. & Tech. Shalimar, Srinagar-191 121 (J&K)	0194-2461258 0191-2462160, 2461551 dnak59@rediffmail.com

Centre	Address of Officer-in-charge	Phone/Fax/Email
KAU, Vellanikkara	Head Department of Olericulture, College of Horticulture, Kerala Agriculture University, Vellanikkara 680 654 (Kerala)	0487-2371918 0487-2370019 psadhankumar@yahoo.co.in
PDKV, Akola	Breeder cum Horticulturist Department of Horticulture, Chilli and Vegetable Research Unit, PKV, Akola-44404 (M.S.)	0674-2444104 0724-2258200 to 2258217 Ext.1141 0724-2258219
VPKAS, Almora	Director VPKAS, Almora-263 601 (Uttranchal)	05962-230208 05962-231539 hedaunirmal_2003@yahoo.co.in
GAU, Anand	Station Head Vegetable Research Station GAU, Anand-388 110	02692-21666, 21871 02692-41520, 41270 rs_mvrs@yahoo.co.in
KKV, Dapoli	Officer Incharge (AICRP-VC) Wakawali, Dapoli, Ratnagiri-415 711 (M.S.)	02358-86932, 86933 visces@rediffmail.com
NHRDF, Nasik	Director NHRDF, Kanda Batata Bhawan, 2954 E, New Mumbai Agra Road, Nasik-422 001 (M.S.)	0253-2592523, 2592531, 2500523, 0253-2596606 nhrdf_nsk@sancharnet.in
HPKV, Palampur	Head Department of Vegetable Crops, HPKV, Palampur, Kangra (H.P.)	01894-30511 pathania_nk@yahoo.co.in
DARL, Pithoragarh	Scientist DRDO Defence Agril. Res. Laboratory, Ministry of Defence, P.B. No. 06 Pithoragarh-262 501 (Uttaranchal)	05964-225564, 224664, 223386, 05964-225564 director@drdarl.ren.nic.in
ICAR, RCNEH, Barapani	Director ICAR, Research Complex of NEH Research, Barapani-793 103 (Meghalaya)	0364-2570363 0364-2570363

Centre	Address of Officer-in-charge	Phone/Fax/Email
NRCOG, Rajguru-nagar	Director NRC for Onion & Garlic Rajgurunagar-410 505, Pune (M.S.)	02135-222026, 222060 02135-224056 nrcog@vsnl.net
CARI, Port Blair	Director Central Agricultural Research Institute, Port Blair-744 101 (A&N)	03192-250436, 251068 03192-233281 drsingh1966@yahoo.com
HARP, Ranchi	Head HARP, Namkum, Ranchi-834 010 (Jharkhand)	0651-2260207, 2260141 0651-2260141 sk_harp@yahoo.co.in
Name of the Officer In-charge and address of seed companies		
Ajeet Seeds	Director Research Ajeet Seeds Limited Gut No. 233, Chittegaon, Taluka Paithan, Aurangabad (Maharshtra)	02431-51445, 51444 02431-51445, 51444, 51508 aurangabad@ajeetseed.com
Ankur Seeds	Research Coordinator Ankur Seeds Pvt. Ltd. 27, New Cotton Market Layout, Nagpur-440 018 (Maharshtra)	0712-2725117, 2726148 0712-723455 makrand@nagpur.dot.net.in
Beejo Sheetal Seeds	Chief Executive Bejo Sheetal Seeds Limited P.B. No. 72, A-2, Mama Chowk Jalna-431 203 (Maharshtra)	02482-232588, 232717 02482-230398 bejosheetal@hotmail.com
Bioseeds Research India Pvt. Ltd.	Deputy Manager (Vegetable Research) Bioseeds Research India Pvt. Ltd. Plot No. 206, H.No 8-2-293/82/A/206, Road No.14, Jubilee Hills, Hyderabad-500 033 (A.P.)	040-23555801-806 040-23555530 sbgi@bioseedindia.com
Century Seeds	Managing Director Century Seeds Pvt. Ltd., Lusa Tower, BA, 22-24, Mangolpuri Industrial Area, Phase-II, New Delhi-110 034	011-7019890, 7017061 011-7017568

Centre	Address of Officer-in-charge	Phone/Fax/Email
Emergent Genetics India Pvt. Ltd.	Director (R&D) 8-2-120/86/5, Avenue:07, Road No. 3, Banjara Hills, Hyderabad-500 034 (A.P.)	040-23555412, 23555419 040-23555393/378 rnd@emergentindia.net
Indo-American Hybrid Seeds	Director (R&D) Indo-American Hybrid Seeds 214, Palika, Bhawan, Sect. XIII, R.K. Puram, New Delhi-110 066	080-2676033 kpvshetty@indamseeds.com
J.K. Agri. Genetics	General Manager (R&D) J.K. Agri. Genetics Limited 1-10-177, 4th Floor, Varun Tower, Begumpet, Hyderabad-500 016 (A.P.)	040-55316858 040-27764943 info@jkseeds.net
Krishidhan Seeds	Director (Research) Krishidhan Seeds Limited, P.O.-92, 34C, Bus Stand Road, Jalna-431 203 (M.S.)	02482-230521, 230809 02482-232328 krishidhanseeds@sify.com
Krishna Seed	Director Krishna Seed Pvt. Ltd. 8/151, Jeoni Mandi, Agra-282 004 (U.P.)	0562-262254
Mahyco Ltd.	Managing Director Mahyco Life Sciences Research Center, Dawalvadi, Post. Box 76, Jalna-431 203 (Maharashtra)	02482-262030 02482-262040 mahycord@bom2.vsnl.net.in
Namdhari Seeds	Managing Director Namdhari Seeds Company 119, 9th Floor Main Road, Ideal Rajarajeshan Nagar, Banglore-560 039 (Karnataka)	
Nirmal Seeds	Director (Research) Nirmal Seeds Limited Bhadgaon Road, Pachora, Jalgaon-424 201 (Maharashtra)	02596-244366, 244396 02596-244045 nspl_jal@sancharnet.in

Centre	Address of Officer-in-charge	Phone/Fax/Email
Nuzi Veedu Seeds	General Manager (R&D) Nuzi Veedu Seeds Limited Processing Plant Survey No. 183 Kompally, Quthbullapur, Mandal Secunderabad-500 014 (Andhra Pradesh)	09440808886
Pandey Beej Bhandar	Manager Pandey Beej Bhandar, Khandari Bazar, Faizabad-224 001 (U.P.)	05278-222810, 222096 05278-228017 pandeyseeds@yahoo.co.in
Pradham Biotech Pvt. Ltd.	General Manager (R&D) Pradham Biotech Pvt. Limited #207, Orange block, Rainbow Complex, Toli Chowki, Hyderabad-500 008 (A.P.)	040 – 23568388 040 - 23568399
Proagro PGS	General Manager (R&D) Proagro PGS India Limited A-305, Ansal Chamber No. 1, 3, Bhikaji Cama Place, New Delhi-110 066	011 - 6191163 011 - 6192084
Seminis Seeds	General Manager (R&D) Seminis Seeds, Gut No. 24, Paithan Road, Chitegaon Aurangabad (Maharashtra)	02431-51468-70 02431-51466
Sreema Seeds	General Manager (R&D) Sreema Seeds Pvt. Ltd. N.H.-5, Naya Chowk Madhupatna, Cuttack-753 010 (Orissa)	0671-2345033, 2345573 0671-2343310 sreemaa_seeds@yahoo. com
Sri Venkates- Hybrid wara Seeds Company	General Manager (R&D) Sri Venkateswara Hybrid Seeds Company, Gosha Hospital Road, Adoni-518 301 (A.P.)	08512 - 25249, 253027 08512-253027

Centre	Address of Officer-in-charge	Phone/Fax/Email
Sungro Seeds	Chief Excutive Sungro Seeds Ltd., 3rd Floor, Sungro Chamber, B.N. Block, Local Shopping Centre, Shalimar Bagh, New Delhi-110 052	011- 7488272, 7471117 011-7470333, 7475525 ceo@sungroseeds.com
Syngenta Seeds	Research Manager Syngenta India Limited Seed Sector, Waogholi, Gate No. 2347, Nagar Road, Taluk, Haveli, Pune (Maharastra)	09120-6688469, 6690228 09120-6688468 asish.patel@syngenta.com
Tokita Seed	General Manager (R&D) Vegetable Breeder Tokita Seed India (P) Limited 360,13th 'A' Main, 80 feet Road, Yelahanka New Town, Bangalore-560 064 (Karnataka)	sham@tokitaindia.com
Tulasi Seeds	General Manager (R&D) Tulasi House, # 6-4-6, Arundelpet 4/5, Guntur (A.P.)	0863-2321374, 2224947 0863-2221161 seeds@tulasigroup.com
Vivaswan Agritech Pvt. Ltd	Director Research Vivaswan Agritech Pvt. Ltd. 125, Bhawani Peth, Gul Ali, Pune-411 042 (Maharashtra)	020-6352945, 6356208
Ventura Crop Sciences	Vice President (R&D) Venntura Crop Sciences Pvt. Ltd. Door No. 270-F, Road No. 10 Jubilee Hills, Hyderabad-500 033 (A.P.)	09490467603 gauri.venntura@gmail. com
VNR Seed	Managing Director VNR, Seeds Pvt. Ltd., 1st Floor, No. 1, Durga College Complex, K.K. Road, Raipur-492 001 (M.P.)	25037846, 25031489 gomchi@vnrseed.com
Zuari Seeds	General Manager (R&D) # 759, Sector-A, III-Phase, Yelahanka New Town, Bangalore-560 064 (Karnataka)	080-8562266, 8460862 080-8562538 greentech@vsno.com

Index